THE ANATOMY and PHYSIOLOGY LEARNING SYSTEM WORKBOOK

Edith J. Applegate
Professor of Science and Mathematics
Kettering College of Medical Arts
Kettering, Ohio

W. B. SAUNDERS COMPANY
A Division of Harcourt Brace & Company
Philadelphia London Toronto Montreal Sydney Tokyo

W. B. SAUNDERS COMPANY
A Division of Harcourt Brace & Company

The Curtis Center
Independence Square West
Philadelphia, Pennsylvania 19106

THE ANATOMY AND PHYSIOLOGY LEARNING SYSTEM Workbook ISBN 0–7216–6638–8

Printed in the United States of America.

PREFACE

The Anatomy and Physiology Learning System includes a textbook, workbook, overhead transparencies, instructor's manual, and a test item file on computer diskette to help teachers present a quality educational experience for students preparing to enter one of the health related professions. It is designed for students enrolled in a one-semester course in human anatomy and physiology. The textbook provides fundamental information and concepts in a manner that encourages learning and understanding. This workbook component of **The Anatomy and Physiology Learning System** provides learning objectives, written exercises that correlate with the objectives, quizzes, medical terminology exercises, and a fun and games page for each chapter. The workbook and textbook together provide a package that facilitates learning, enhances understanding, relates information to medical applications, and enables the students to apply their knowledge to further study within their chosen health related professions.

First time students in anatomy and physiology, particularly those with limited science background, sometimes find the concepts and terminology difficult and possibly intimidating. Often these students also lack the study skills necessary to master difficult themes. They need step-by-step help and guidance through the learning process. This workbook provides this assistance to learning and understanding. By using the workbook, the students are able to more actively participate in their own learning. It also provides a focus for small group study with their peers. Not only do these activities benefit the student, but they also benefit the teacher. As students become more involved in their own learning process, they tend to rely less on individualized or private tutoring by the teacher. By spending less time "tutoring," the teacher has more time to prepare informative and stimulating lectures. As the classes become more interesting, the students become more excited about learning, and everybody wins!

The workbook is designed to have several possible uses. The pages are perforated so that any part or all of the exercises may be torn out and turned in for an assignment. The answers to the exercises are not included in the workbook, but they are in the instructor's manual. This gives the teacher flexibility in using the workbook. If some of the workbook exercises are used for assignments, the answers are available in the instructor's manual for grading purposes. If other exercises are used for supplemental self-study, those answers may be posted for the students to self-check. Specific workbook and instructor's manual features are described below.

WORKBOOK FEATURES

- **Chapter learning objectives** are organized according to textbook chapter outline headings. These objectives help the student to focus on the important items presented in each chapter.
- **Learning exercises** are organized according to the chapter outlines and objectives. Five or more pages of exercises, including short answer, matching, and diagrams to label are provided for each chapter.
- **Self-quizzes** give students an opportunity to check on their progress and understanding. There is a self quiz for each chapter.
- **Terminology exercises** give the student an opportunity to create and define words that relate to the topics presented in each chapter by using prefixes, suffixes, and word roots. This is excellent practice for their study in other courses relating to their chosen health related career.
- **Fun and games** is a lighter, more relaxed approach to the topics in each chapter. The final page of each workbook chapter is some type of word puzzle that incorporates words and concepts from the chapter. The workbook includes a variety of puzzles that provide reinforcement and show that learning can be fun.

ADDITIONAL MATERIAL FOR THE INSTRUCTOR

- **Instructor's Manual** is provided to adopters of **The Anatomy and Physiology Learning System**. This manual includes the answers to the review questions at the end of each chapter in the textbook and the answers for the learning exercises, quizzes, terminology exercises, and puzzles in the workbook. The manual also includes a collection of test questions with answers for each chapter.
- **Overhead Transparencies** are provided to adopters for use in lecture presentations. Fifty full-color illustrations from the textbook have been selected for this transparency set. They are certain to add interest and color to your lectures.
- **Computerized Test Bank** is provided to adopters for use in preparing tests and quizzes. This computer diskette package contains a minimum of fifty questions for each chapter to give a total of at least one thousand questions.

NOTE TO STUDENTS

Dear Student,

I have prepared this workbook with you in mind. It contains exercises to help you organize and learn anatomy and physiology. These exercises have different formats--questions to answer, statements to complete, paragraphs in which you fill in missing words, matching exercises, and diagrams to label color. Many of the exercises ask you to write key words because this helps you learn to spell the words correctly. A small set of colored pencils will be useful because many of the labeling exercises suggest that you color code some of the structures on the diagram with the appropriate label.

There is a quiz provided for each chapter. You can use this as a measuring tool to see how well you are doing. Scientific terminology is important in anatomy and physiology and also in all health-related fields. For this reason, I have included a page of terminology exercises in each chapter. These will give you practice in defining new words and in building words from word parts that you know. As you become more familiar with terminology, A & P becomes easier.

To close each chapter in the workbook, I have prepared some type of word puzzle that relates to the topics in the chapter. These are for those moments when you don't feel like studying, but know you should and feel guilty about not doing it. Maybe the puzzles will get you in the mood. I had a lot of fun preparing the **Fun and Games** pages and I hope you have fun doing them. My premise is that you can learn and have fun at the same time.

Carrying this workbook to and from class may help your cardiovascular system, but it won't help you learn anatomy and physiology unless you actually **use** it. You learn better when you are actively involved in the process and this workbook provides a means for your participation and involvement in your learning process. It offers an alternative method of study after you have studied your notes and textbook. The workbook also provides a focus for small group study with your peers. This is a WORK book. Use it. You'll be glad you did.

I hope you enjoy your study of the human body. I would like to hear from you and have your comments and suggestions about ways to improve **The Anatomy and Physiology Learning System.** In closing, I offer my best wishes for success this course, in your selected curriculum, and in your chosen career.

Sincerely,

Edith Applegate

Edith Applegate
Kettering College of Medical Arts
Department of Biological Science
3737 Southern Boulevard
Kettering, OH 45429-1299

1 Introduction to Anatomy and Physiology

☞ Chapter Outline/Objectives

Anatomy and Physiology

1. Define the terms anatomy and physiology and discuss the relationship between the two areas of study.

Levels of Organization

2. List the six levels of organization within the human body.

Organ Systems

3. Name the eleven organ systems of the body and briefly describe the major role of each one.

Life Processes

4. List and define ten life processes in the human body.
5. List five physical environmental factors necessary for survival of the individual.

Homeostasis

6. Discuss the concept of homeostasis.
7. Distinguish between negative feedback mechanisms and positive feedback mechanisms.

Anatomical Terms

8. Describe the four criteria that are used to describe the anatomic position.
9. Distinguish between the terms superior and inferior, anterior and posterior, medial and lateral, proximal and distal, superficial and deep, visceral and parietal.
10. Describe sagittal, midsagittal, transverse, and frontal planes.
11. Distinguish between the dorsal body cavity and the ventral body cavity, and list the subdivisions of each one.
12. Locate the nine abdominal regions.
13. Distinguish between the axial and appendicular portions of the body.
14. Use anatomic terms that relate to specific body areas.

☞ Learning Exercises

Anatomy and Physiology (Objective 1)

1. The study of morphology or structure of organisms is called ____________________.

2. ____________________ is the scientific study of body functions.

3. Anatomy and physiology are interrelated because the structure of a body part influences its ____________________ and function has an effect on ____________________.

4. Identify the specialty areas of anatomy and physiology that are described below by writing the name of the specialty area on the line preceding the description.

______________________ Study of external features

______________________ Study of cellular structure

______________________ Study of prenatal development

______________________ Study of the body's defense against disease

______________________ Study of drug action in the body

______________________ Study of structural and functional changes associated with disease

Levels of Organization (Objective 2)

1. The six levels of organization in the body, in sequence from the simplest to the most complex, are ______________________, ______________________, ______________________, ______________________, ______________________, ______________________.

2. The basic living unit of all organisms is the ______________________.

3. A ______________________ is a collection of cells with similar structure and function.

Organ Systems (Objective 3)

Write the name of the organ system that corresponds to each of the descriptions.

Cardiovascular	Lymphatic	Respiratory
Digestive	Muscular	Skeletal
Endocrine	Nervous	Urinary
Integumentary	Reproductive	

1. ______________________ Consists of the skin, hair, and sweat glands

2. ______________________ Bones and ligaments

3. ______________________ Processes food into usable molecules

4. ______________________ Trachea, bronchi, and lungs

5. ______________________ Removes nitrogenous wastes from the blood

6. ______________________ Glands that secrete hormones

7. ______________________ Cleanses lymph and returns it to the blood

8. ______________________ Brain, spinal cord, nerves, and sense receptors

9. ______________________ Blood, heart, and blood vessels

10. ______________________ Part of body's defense system

11. ______________________ Esophagus, stomach, liver, and pancreas

12. ______________________ Transmits impulses to coordinate body activities

13. ______________________ Chemical messengers that regulate body activities

14. ______________________ Transports nutrients, hormones, and oxygen

15. ______________________ Regulates fluid and chemical content of the body

16. ______________________ Produces movement and maintains posture

17. ______________________ Tonsils, spleen, lymph nodes, and thymus

18. ______________________ Protective covering of the body

19. ______________________ Forms the framework of the body

20. ______________________ Ovaries and testes

Life Processes (Objectives 4 and 5)

1. List ten life processes that dinstinguish living organisms from non-living forms.

2. ______________________ is the phase of metabolism in which complex substances are broken down into simpler ones.

3. Define anabolism.

4. List five physical factors from the environment that are essential to human life.

Homeostasis (Objectives 6 and 7)

Write the term that is defined by each of the following phrases.

1. ______________________ Maintenance of a relatively stable internal environment
2. ______________________ Any condition that disrupts homeostasis
3. ______________________ Action that has an effect opposite to a deviation from normal; action to maintain homeostasis
4. ______________________ Mechanisms that stimulate or amplify changes

Anatomical Terms (Objectives 8-14)

1. Describe the anatomical position by stating the position of each of the following parts:

   Body is ______________________ Face is ______________________

   Arms are ______________________ Palms are ______________________

   Feet and toes are ______________________

2. Provide the directional term that correctly completes each statement.

   The nose is ______________________ to the mouth.

   The elbow is ______________________ to the wrist.

   Muscles are ______________________ to the skin.

   The heart is ______________________, or in front of, the vertebral column.

   The ______________________ pericardium covers the heart.

3. Name the plane that:

   ______________________ Divides the body into right and left halves.

   ______________________ Divides the body into superior and inferior portions.

   ______________________ Divides the body into anterior and posterior portions.

   ______________________ Divides the body into right and left portions.

4. Name the most specific body cavity that is described by each of the following phrases:

______________________	Contains the cranial and spinal cavities
______________________	Contains the thoracic and abdominopelvic cavities
______________________	Contains the brain
______________________	Contains the heart and lungs
______________________	Contains the liver, stomach, and spleen
______________________	Contains the spinal cord
______________________	Contains the urinary bladder and rectum

5. Label the nine regions of the abdomen indicated on the following diagram.

A	D	F
G	B	E
I	H	C

A. ______________________ F. ______________________

B. ______________________ G. ______________________

C. ______________________ H. ______________________

D. ______________________ I. ______________________

E. ______________________

6. Label the body regions indicated in the following figures.

A. ______________________

B. ______________________

C. ______________________

D. ______________________

E. ______________________

F. ______________________

G. ______________________

H. ______________________

I. ______________________

J. ______________________

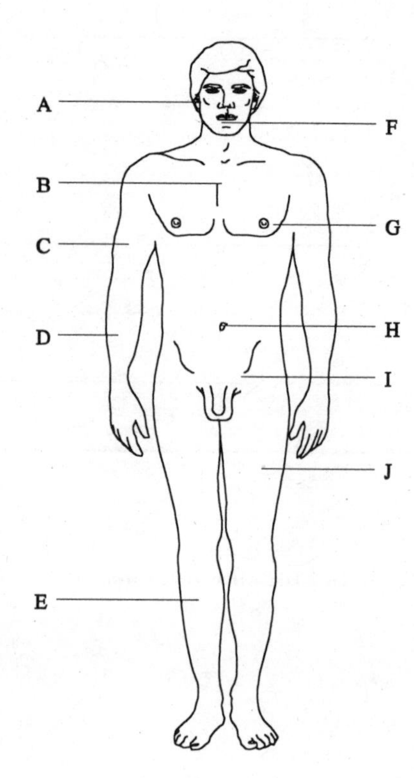

K. ______________________

L. ______________________

M. ______________________

N. ______________________

O. ______________________

P. ______________________

Q. ______________________

R. ______________________

S. ______________________

T. ______________________

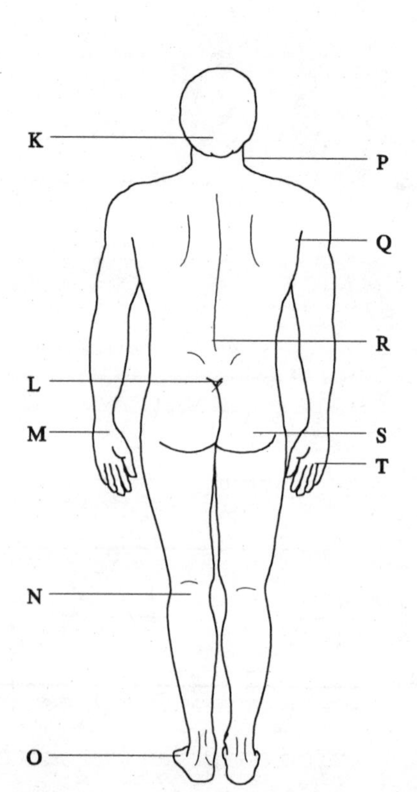

☞ Chapter Self-Quiz

1. ______________________________ is the scientific study of how the parts of the body work.

2. The smallest unit of organization that is <u>living</u> is (a) chemical; (b) cell; (c) tissue; (d) organ; (e) system

3. Which of the following does <u>not</u> represent a correct grouping of organ system/part of system/function? (a) integumentary/skin/cover and protect; (b) endocrine/ductless glands/regulate metabolic activity; (c) respiratory/lungs/exchange of gases; (d) lymphatic/heart/defense against disease; (e) urinary/kidney/excrete metabolic wastes

4. The "breaking-down" part of metabolism is called ______________________________.

5. Which of the following correctly lists three physical factors from the environment that are necessary to sustain human life? (a) oxygen, metabolism, respiration; (b) water, organization, excretion; (c) oxygen, water, anabolism; (d) oxygen, water, pressure; (e) reproduction, respiration, nutrients

6. Which one of the following does <u>not</u> pertain to a negative feedback mechanism? (a) helps to maintain homeostasis; (b) when blood pressure decreases, the heart beats faster to increase blood pressure; (c) responsible for increased sweating when air temperature is higher than body temperature; (d) when blood sugar level decreases, the hunger center in the brain is stimulated; (e) increases deviations from normal

7. In anatomical position, (a) your body is erect, arms are behind your back; (b) your eyes are facing the same direction as your palms; (c) you are sitting down with feet forward; (d) feet are forward and palms are in the opposite direction; (e) arms are at your side and palms are facing the opposite direction from your eyes

8. Which of the following means closer to a point of attachment or origin? (a) superficial; (b) anterior; (c) proximal; (d) medial; (e) superior

9. The plane that divides the body into anterior and posterior parts is the ______________ plane

10. The region that is in the midline, superior to the umbilical region, is the (a) hypogastric; (b) hypochondriac; (c) lumbar; (d) epigastric; (e) iliac

11. Place an X before each of the following that pertains to the ventral body cavity.

_____ Spinal cord

_____ Heart

_____ Thoracic cavity

_____ Small intestines

_____ Cranial cavity

12. The appendicular portion of the body includes the (a) head and neck; (b) arms and legs; (c) thorax and abdomen; (d) brain and spinal cord; (e) dorsal and ventral cavities

13. Match the following terms with the appropriate body region.

_______ Skull	A. antecubital
_______ Buttock region	B. axillary
_______ Armpit area	C. cervical
_______ Chest region	D. cranial
_______ Area behind the knee	E. gluteal
_______ Middle region of abdomen	F. oral
_______ Anterior midline of thorax	G. pectoral
_______ Space in front of the elbow	H. popliteal
_______ Mouth	I. sternal
_______ Neck region	J. umbilical

☞ Terminology Exercises

WORD PART	MEANING
al-	pertaining to
ana-	apart
cardi-	heart
dors-	back
epi-	upon, above
gastr-	stomach
homeo-	alike, same
integ-	a covering
-ism	process of
-itis	inflammation
-logy	study of, science of

WORD PART	MEANING
metabol-	change
path-	disease
pelv-	basin
physi-	nature, function
proxim-	nearest
skelet-	a dried, hard body
-sta-	to control, staying
-tom-	to cut
vas-	vessel
viscer-	internal organs
-y	process, condition

Use word parts given above to form words that have the following definitions.

______________________________ Pertaining to the back

______________________________ Inflammation of the stomach

______________________________ Study of the heart

______________________________ Staying the same

______________________________ The process of change

Using the definitions of word parts given above, define the following words.

Proximal ______________________________

Visceral ______________________________

Epigastric ______________________________

Skeletal ______________________________

Pathology ______________________________

Match each of the following definitions with the correct word.

__________	Process of cutting apart	A. physiology
__________	The skin or covering of the body	B. pelvis
__________	The study of function	C. cardiovascular
__________	Structure shaped like a basin	D. anatomy
__________	Pertaining to the heart and blood vessels	E. integument

☞ Fun and Games

The definitions for thirty words are given below. Determine the correct term for each definition, write the term in the space provided in the word list, then find the terms in the word search puzzle. Words may be horizontal, vertical, or diagonal and may read forward or backward

Word Search Puzzle

```
L A M I X O R P G Q E X I F E W U X J Z
A T L M I F R A X C B Y R O R L B H E F
E R Y E M S C W A A R L S A O A Z C N R
T O T T U L I T N O A A U B I P T E S O
I I K A Y B A A T T G C P D R R G L O N
L R E B M B B A E I B C B O E A L L T T
P E J O O O R R T S Z U G M T C V S Y A
O P I L L I A T M S S B V I S C E R A L
P U I I P L A Z H U H J V N O Y X M A O
T S S S Z L K N T E F E L O P Q Y S U L
M M E M Q D F Y T S F A I P R G Z M I E
Y R A T N E M U G E T N I E O Y B S E R
U E A S M O O C E I B V P L U I S C V O
F R C O T Q I D P G U R O V L E K P I S
E R R A S C B I N C G I A I E V E D T S
P A N H A A C J J N S S C C S Q L B S E
L A J R C C B X M Y P A E L H I E T E R
S Z O K O W E P H A L S B B S I T S G T
X H H V C C C P S H O M E O S T A S I S
T V R B B K L N J M J O B F A F L L D O
```

Definitions

1. Large ventral body cavity inferior to diaphragm
2. Building phase of metabolism
3. Study of structure
4. Region from elbow to wrist
5. Cheek region
6. Region of the wrist
7. Breaking down phase of metabolism
8. Smallest living unit
9. Body system that includes the esophagus and stomach
10. Thigh region
11. Plane dividing body into anterior and posterior
12. Stable internal environment
13. Body system that includes the skin and sweat glands
14. Toward the side
15. All the chemical reactions that occur in the body
16. Normally maintains homeostasis
17. Lower portion of back of head
18. Study of function
19. Area behind the knee
20. Toward the back, dorsal
21. Closer to the attachment
22. Body system that includes the bronchi and lungs
23. Plane that divides the body into right and left portions
24. System that includes bones
25. A condition that disrupts homeostasis
26. Above another portion
27. Ventral body cavity that contains the heart and lungs
28. Groups of cells with similar structure and function
29. Middle region of abdomen
30. Pertains to the internal organs

Word List

1. ____________
2. ____________
3. ____________
4. ____________
5. ____________
6. ____________
7. ____________
8. ____________
9. ____________
10. ____________
11. ____________
12. ____________
13. ____________
14. ____________
15. ____________
16. ____________
17. ____________
18. ____________
19. ____________
20. ____________
21. ____________
22. ____________
23. ____________
24. ____________
25. ____________
26. ____________
27. ____________
28. ____________
29. ____________
30. ____________

2 Chemistry, Matter, and Life

☞ Chapter Outline/Objectives

Elements

1. Define matter.
2. Define an element.
3. Use chemical symbols to identify elements.

Structure of Atoms

4. Differentiate between protons, neutrons, and electrons, and tell where each one is located.
5. Draw a simplified diagram that illustrates the structure of an atom.
6. Distinguish between atomic number and mass number of an element.
7. Describe the electron arrangement that makes an atom most stable.

Chemical Bonds

8. Describe the difference between ionic bonds, covalent bonds, and hydrogen bonds.
9. Distinguish between cations and anions.

Compounds and Molecules

10. Describe the relationship between atoms, molecules, and compounds.
11. Interpret molecular formulas for compounds.

Chemical Reactions

12. Identify the reactants and products in a chemical equation.
13. Describe and illustrate four types of chemical reactions.
14. Compare exergonic and endergonic reactions.
15. Discuss five factors that influence the rate of chemical reactions.
16. Explain what is meant by a reversible reaction.

Mixtures, Solutions, and Suspensions

17. Distinguish between mixtures, solutions, and suspensions.

Acids, Bases, and Buffers

18. Define the term "electrolyte."
19. Describe what makes an acid or a base and what happens when they react.
20. Discuss the concepts of pH and buffers.

Organic Compounds

21. Describe the five major groups of organic compounds that are important to the human body.

☞ Learning Exercises

Elements (Objectives 1-3)

1. Matter is defined as ____

   ____.

2. An element is defined as ____

   ____.

3. Write the chemical symbol for each of the following elements.

   ______ Hydrogen ______ Oxygen ______ Magnesium

   ______ Potassium ______ Carbon ______ Phosphorus

4. Identify the element that is represented by each of the following symbols.

   ______________________ Na ______________________ Ca

   ______________________ Cl ______________________ N

   ______________________ Fe ______________________ S

Structure of Atoms (Objectives 4-7)

1. Draw a simple diagram that illustrates the structure of an oxygen atom, which has an atomic number = 8 and a mass number = 16.

2. Fill in the information that is missing from the following table.

Particle	Location	Charge	Mass
		0	
			Negligible
Proton			

3. How many protons, neutrons, and electrons are in an atom of potassium, which has an atomic number = 19 and a mass number = 39?

   ______ Protons ______ Neutrons ______ Electrons

4. The most stable atoms have __________ electrons in their highest energy level.

Chemical Bonds (Objectives 8 and 9)

Match each of the following terms with the correct definition below.

A. Ionic bond
B. Covalent bond
C. Hydrogen bond
D. Anion
E. Cation

______ Intermolecular bond

______ Ion with a negative charge

______ Ion with a positive charge

______ Bond between anions and cations

______ Bond in which electrons are shared

______ Bond between polar covalent molecules

Compounds and Molecules (Objectives 10 and 11)

1. Indicate whether each of the following refers to an atom, a molecule, or a compound.

______________________ Smallest unit of an element that retains the properties of that element

______________________ Formed when two or more atoms are held together by chemical bonds

______________________ Formed when two or more different atoms are chemically combined in a definite ratio

______________________ Smallest unit of a compound that retains the properties of that compound

2. The molecular formula for calcium carbonate, one of the mineral salts in bone, is $CaCO_3$. Name the elements in this compound and tell how many atoms of each element are present.

______________ ______________ ______________

Chemical Reactions (Objectives 12-16)

1. For each example of an equation representing a chemical reaction, indicate whether it is synthesis, decomposition, single replacement, or double replacement and write the formulas for the reactants and products.

Equation: $H_2CO_3 \rightarrow H_2O + CO_2$
carbonic acid

Type of reaction:

Reactants:

Products:

Equation: $N_2 + 3H_2 \rightarrow 2NH_3$

Type of reaction:

Reactants:

Products:

Equation: $MgCl_2 + 2NaOH \rightarrow Mg(OH)_2 + 2NaCl$
milk of magnesia

Type of reaction:

Reactants:

Products:

Equation: $C_7H_6O_3 + C_2H_4O_2 \rightarrow C_9H_8O_4 + H_2O$
salicylic acid, acetic acid, aspirin

Type of reaction:

Reactants:

Products:

2. In ______________________________ reactions, there is more energy stored in the reactants than in the products. Energy is released in these reactions. Reactions that need an input of energy are called ______________________________ reactions.

3. For each of the following changes, indicate whether the reaction rate will increase (I) or decrease (D).

________	Grind up the reactants	______	Add a catalyst
________	Dilute one reactant	______	Use more concentrated solutions
________	Cool the reaction mixture	______	Increase the temperature

4. What is the meaning of the double arrow in the following equation?

$CO_2 + H_2O \rightleftarrows H^+ + HCO_3^-$

Mixtures, Solutions, and Suspensions (Objective 17)

1. When salt is dissolved in water, the water is called the (a) solution; (b) solute; (c) solvent; (d) liquid; (e) dialysate.

2. Solutions (a) are clear; (b) have a fixed composition; (c) settle when left standing; (d) must be separated by chemical means.

3. Indicate whether each of the following combinations is most accurately described as a mixture, solution, or suspension.

___________ Sugar and salt

___________ Sugar and water

___________ Sand and water

___________ Blood cells and plasma

___________ Cytoplasm of the cell

Acids, Bases, and Buffers (Objectives 18-20)

1. Substances that form ions in solution are called ______________________________.

2. Indicate whether each of the following phrases refers to an acid or to a base.

_______ Accepts hydrogen ions

_______ Has a sour taste

_______ Has a pH of 3.5

_______ Donates protons

______ Reacts with a buffer to form a weak acid

______ Reacts with OH^- ions to form water

______ Has a slippery, soapy feeling

______ Has a pH of 8.7

3. Acetic acid ($HC_2H_3O_2$) and sodium acetate ($NaC_2H_3O_2$) are members of a buffer pair.

   Which member of the buffer pair reacts to neutralize sodium hydroxide (NaOH) and what neutral product is formed?

   Which member of the buffer pair reacts to neutralize hydrochloric acid (HCl) and what neutral product is formed?

Organic Compounds (Objective 21)

1. Carbohydrates contain the elements ______________, ____________, and __________.
2. Three important hexose monosaccharides are ______________________________, ______________________________, and ______________________________.
3. When two monosaccharides are joined by chemical bonds, the resulting molecule is called a ______________________________.
4. Starch, cellulose, and glycogen are examples of a group of carbohydrates called ______________________________.

5. Write the term that is described by each of the following phrases.

_________________________ Element, in addition to C, H, and O, found in all proteins

_________________________ Building blocks of proteins

_________________________ Amino acids that cannot be synthesized in the body

6. Name the following items that pertain to lipids.

_________________________ The building blocks of triglycerides

_________________________ Fatty acids that contain all single covalent bonds

_________________________ Lipids that are an important component of cell membranes

7. Write the term that matches each of the following phrases in the spaces at the left.

_________________________ Forms the genetic material of the cell

_________________________ Single-stranded nucleic acid involved in protein synthesis

_________________________ Building units of nucleic acids

_________________________ Sugar in DNA molecules

_________________________ Double-stranded nucleic acid

_________________________ Five elements in nucleic acids

_________________________ Nucleic acid that contains uracil

_________________________ High energy compound that contains three phosphate groups

8. Match the following substances with the correct class of organic compounds.

A. Carbohydrates B. Lipids C. Proteins D. Nucleic acids

________	Glucose	________	Amino acids
________	Steroids	________	Nucleotides
________	Glycerol	________	Glycogen
________	Hemoglobin	________	RNA
________	Disaccharides	________	Triglycerides

☞ Chapter Self-Quiz

1. Write the name of the element represented by each of the following symbols:

(a) C ____________ (c) Cl____________ (e) Cu____________

(b) P ____________ (d) Fe____________ (f) O ____________

2. A certain neutral atom has a mass number of 35 and has 17 electrons.

(a) What is the atomic number of this atom?

(b) How many neutrons are there in the atom?

(c) How many protons are there in the atom?

3. Match the terms on the right with the descriptions and definitions on the left.

______ Equals the number of protons plus the number of neutrons

______ Contained in shells or energy levels surrounding the nucleus

______ Number of these particles equals the number of electrons in a neutral atom

______ Determined by the number of protons in the atom

______ Have the same mass as protons

A. atomic number

B. electrons

C. mass number

D. neutrons

E. protons

4. Given the following reaction:

$$Zn + CuSO_4 \rightarrow ZnSO_4 + Cu$$

(a) Name the four elements in this reaction.

(b) Write the chemical formulas for the reactants.

(c) How many atoms of oxygen are in the products?

(d) What type of reaction is this?

5. Indicate whether each of the following pertains to an acid or to a base. Use A = acid and B = base.

______ Proton donor

______ pH = 8.2

______ Accepts hydrogen ions

______ pH = 2.5

______ Accepts protons

______ pH = 6.5

______ Donates hydrogen ions

6. Given the following neutralization reaction:

$NaOH + HCl \rightarrow NaCl + H_2O$

Identify the **acid** __________, **base** ____________, and **salt** __________.

7. Assume that a reaction is proceeding at a given rate. Indicate how each of the following will affect the reaction rate. Use F = faster rate and S = slower rate.

_____ Remove some of the product

_____ Add an appropriate catalyst

_____ Decrease the temperature

_____ Break the reactants up into smaller particles

_____ Increase the concentration of the reactants

8. Identify each of the following as a monosaccharide (M), disaccharide (D), or polysaccharide (P).

_______ Cellulose

_______ Fructose

_______ Galactose

_______ Glucose

_______ Glycogen

_______ Lactose

_______ Maltose

_______ Sucrose

9. Classify each of the following as pertaining to carbohydrates (C), lipids (L), proteins (P), or nucleic acids (N).

_______ Adenine, cytosine

_______ Amino acids

_______ Cholesterol

_______ Monosaccharides

_______ Nucleotides

_______ Peptide bonds

_______ Starch

_______ Triglycerides

10. Which of the following is not true about nucleic acids? (a) DNA consists of two strands; (b) both DNA and RNA contain nucleotides; (c) RNA functions in protein synthesis; (d) both DNA and RNA contain a five-carbon sugar; (e) both DNA and RNA contain the nitrogenous base thymine

☞ Terminology Exercises

WORD PART	MEANING	WORD PART	MEANING
alkal-	basic	lact-	milk
carb/o-	charcoal, coal, carbon	lip/o-	fat
di-	two	-lys	to take apart
end-	within, inner	mono-	one
erg-	work, energy	-ose	sugar
ex-	out of, away from	oxy-	oxygen
-genesis	to form, produce	pent-	five
hex-	six	poly-	many
hydro-	water	sacchar-	sugar, sweet
-ide	pertaining to	tri-	three

Use word parts given above to form words that have the following definitions.

_______________________________ Many sugars

_______________________________ Produce fat

_______________________________ Molecule with two phosphates

_______________________________ Taking energy away from

_______________________________ Pertaining to two oxygens

Using the definitions of word parts given above, define the following words.

Triphosphate _______________________________

Endergonic _______________________________

Disaccharide _______________________________

Lactose _______________________________

Hydrolysis _______________________________

Match each of the following definitions with the correct word.

__________ Forming glycogen

__________ Less oxygen than ribose

__________ Basic, pH greater than 7

__________ Pertaining to carbon and water

__________ Breaking down fat

A. alkaline
B. carbohydrate
C. deoxyribose
D. glycogenesis
E. lipolysis

☞ Fun and Games

Determine the words that fit the clues in the left column, then write the words in the boxes in the right column, one letter per box. The circled letters form a scrambled word or phrase. Unscramble the letters to create the word or phrase that matches the final clue.

A. Smallest unit of a compound

Positively charged subatomic particle

Building block of a protein

Accepts protons

Product in a neutralization reaction

Speeds up a chemical reaction

One of the elements in water

A complex carbohydrate

Final clue: High energy compound ______________________

B. An element in water

Sugar in RNA

Element always present in proteins

Group of compounds that includes fats

A base in nucleic acids

Positively charged ion

Negatively charged subatomic particle

Resists change in pH

Fat with three fatty acids

Final Clue: Genetic material of the cell ______________________

3 Cell Structure and Function

☞ Chapter Outline/Objectives

Structure of the Generalized Cell

1. Explain what is meant by a "generalized cell."
2. Describe the composition of the cell membrane.
3. List five functions of the proteins in the cell membrane.
4. Describe the cytoplasm.
5. Describe the components of the nucleus and state the function of each one.
6. Describe each of the cytoplasmic organelles and state the function of each one.
7. Characterize the cytoskeleton.
8. Relate the structure, location, and function of the centrioles.
9. Distinguish between cilia and flagella on the basis of structure and function.
10. Identify the parts of a generalized cell.

Cell Functions

11. Explain how the cell membrane regulates the composition of the cytoplasm.
12. Describe the process of diffusion and give a physiologic example.
13. Distinguish between simple diffusion and facilitated diffusion.
14. Explain the difference between simple diffusion and osmosis.
15. Explain the significance of isotonic, hypertonic, and hypotonic solutions.
16. Describe the mechanics of filtration and cite two physiologic examples.
17. Distinguish between active transport and other types of membrane transport.
18. Distinguish between endocytosis and exocytosis and give examples.
19. Name the phases of a typical cell cycle and describe the events that occur in each phase.
20. Explain the difference between mitosis and meiosis.
21. Explain what constitutes a gene.
22. Describe the process of DNA replication.
23. Explain how DNA in the nucleus regulates protein synthesis in the cytoplasm by relating the events that occur during protein synthesis.
24. Define the terms transcription and translation as they pertain to protein synthesis.

☞ Learning Exercises

Structure of the Generalized Cell (Objectives 1-10)

1. List five functions of the various proteins that are found in the cell membrane.

   a.

   b.

   c.

   d.

   e.

2. Identify the parts of the generalized cell in the following diagram by placing the correct letter in the space provided by the listed labels. Then select different colors for each structure and use them to color the coding circles and the corresponding structures in the illustration.

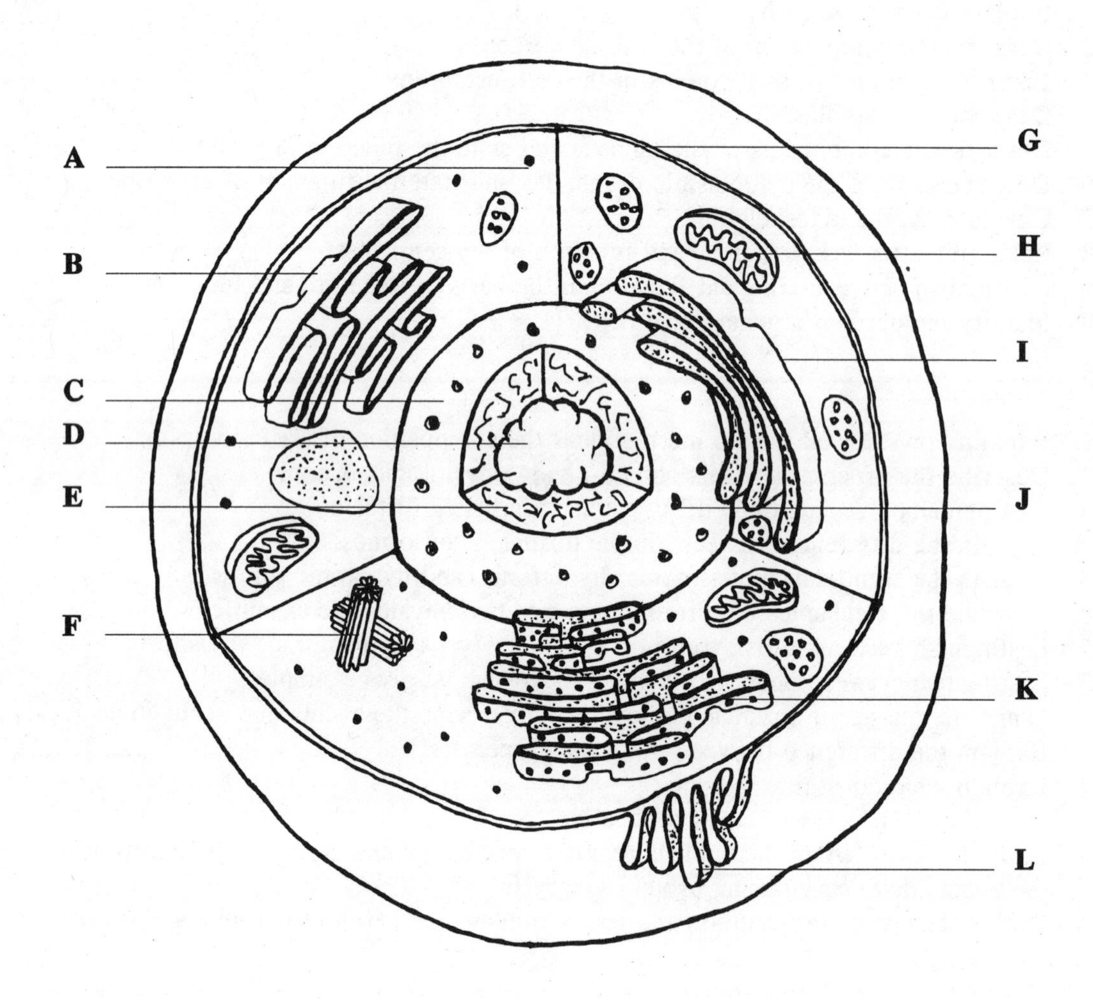

____ Cell membrane ○	____ Golgi apparatus ○	____ Nucleus ○
____ Centriole ○	____ Mitochondria ○	____ Ribosome ○
____ Chromatin ○	____ Nuclear pore ○	____ RER ○
____ Cilia ○	____ Nucleolus ○	____ SER ○

3. Name the cellular component described by each of the following phrases.

_________________________ Contains enzymes for the production of ATP

_________________________ Contains digestive enzymes

_________________________ Prepares products for exocytosis

_________________________ Granules of RNA in the cytoplasm

_________________________ Located in the centrosome

_________________________ Propel substances across the surface of the cell

_________________________ Membranous transport system within the cell

_________________________ Long strands of DNA within the nucleus

_________________________ Function to move the cell

_________________________ Dark structure that contains RNA in the nucleus

_________________________ Semiliquid around the organelles

_________________________ Controls flow of materials into and out of the cell

_________________________ Organelle that controls cell functions

_________________________ Form the cytoskeleton (2 answers)

____

Cell Functions (Objectives 11-24)

1. Indicate whether each of the following refers to simple diffusion, facilitated diffusion, osmosis, or filtration.

_________________________ Moves water through a selectively permeable membrane

_________________________ Movement of gases in the lungs

_________________________ Requires a carrier molecule

2. Identify each of the following transport mechanisms as active transport, passive transport, endocytosis, or exocytosis.

_________________________ Utilizes ATP and a carrier molecule

_________________________ Phagocytosis

_________________________ Osmosis

_________________________ Releases secretory products through the cell membrane

_________________________ Transports ions from low to high concentrations

_________________________ Facilitated diffusion

_________________________ Incorporates fluid droplets into the cytoplasm

3. The accompanying diagram represents two solutions separated by a selectively permeable membrane. Answer the following questions about these solutions.

Which solution is hypertonic?

Which side will increase in volume because of osmosis?

Solution A 5%	Solution B 1%

4. Identify the phase of mitosis that is represented by each of the following illustrations.

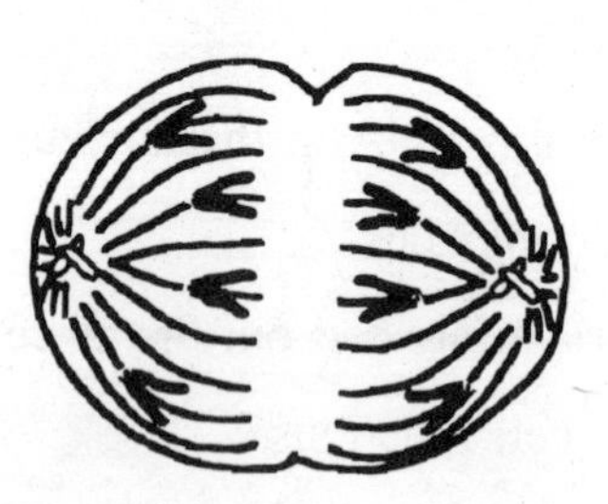

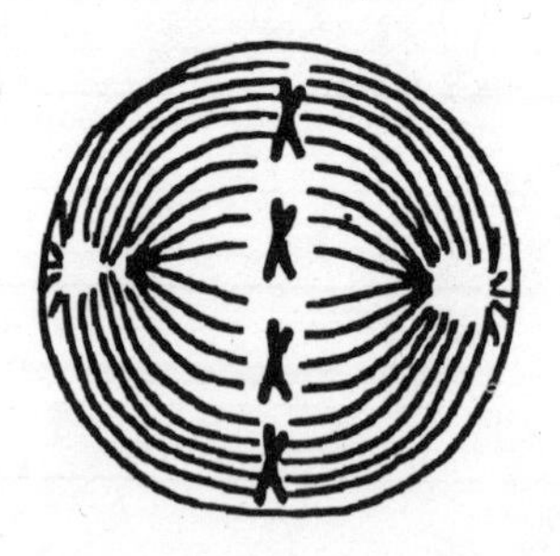

5. Compare mitosis and meiosis by answering the questions in the following table.

	Mitosis	Meiosis
In what type of cell does each occur?		
How many new cells are produced in each completed division?		
If the original cell has 46 chromosomes (23 pair), how many are in each new cell?		

6. The following is a sequence of bases in a portion of the gene that codes for the human hormone oxytocin.

A = adenine C = cytosine G = guanine T = thymine U = uracil

T A C A C A A T G T A A G T T T T G

Write out the sequence of bases on the corresponding mRNA molecule and mark off the codons on this molecule.

Write out the sequence of bases on the six anticodons on the tRNA molecules.

☞ Chapter Self-Quiz

1. Match each description with the correct cell component.

_______	In the cytoplasm; functions in protein synthesis	A.	centrioles
_______	Contains chromatin	B.	cilia
_______	Dense area of RNA in the nucleus	C.	endoplasmic reticulum
_______	Functions in the transport of proteins and lipids	D.	flagella
_______	Protein organelle that functions in cell division	E.	Golgi apparatus
_______	Modifies and prepares substances for secretion	F.	lysosomes
_______	Long, whiplike structures that propel the cell	G.	mitochondria
_______	Contain digestive enzymes	H.	nucleolus
_______	Contain enzymes for making ATP	I.	nucleus
_______	Move substances across the surface of the cell	J.	ribosomes

2. Which one of the following does not move solutes from a region of higher concentration to a region of lower concentration? (a) simple diffusion; (b) facilitated diffusion; (c) osmosis

3. Which one of the following uses cellular energy? (a) simple diffusion; (b) active transport; (c) facilitated diffusion; (d) osmosis

4. When a red blood cell is placed in a hypotonic solution, the cell will (a) shrink; (b) swell up; (c) crenate; (d) more than one of the responses are correct

5. ________________________ is described as "cell drinking" because it is taking in liquid droplets.

6. Match the following descriptions with the correct phase of the cell cycle.

_______	Period of growth and metabolism	A. anaphase
_______	Chromatin shortens and thickens to form chromosomes	B. interphase
		C. metaphase
_______	Chromosomes align themselves along the center of the cell	D. prophase
_______	Nuclear membrane disappears	E. telophase
_______	Chromosomes migrate to the end of the cell	
_______	Nucleolus disappears	
_______	Cytokinesis occurs	

7. If a cell has 46 chromosomes, (a) there will be 4 cells, each with 23 chromosomes after mitosis; (b) there will be 4 cells, each with 46 chromosomes after mitosis; (c) there will be 2 cells, each with 23 chromosomes after mitosis; (d) there will be 2 cells, each with 46 chromosomes after mitosis.

8. In DNA replication, (a) adenine always pairs with thymine; (b) adenine always pairs with cytosine; (c) adenine always pairs with guanine; (d) cytosine always pairs with thymine; (e) guanine always pairs with thymine.

9. During transcription, (a) DNA is replicated; (b) tRNA transfers an amino acid to a ribosome; (c) a codon on mRNA pairs with an anticodon on tRNA; (d) genetic information is transferred from DNA to mRNA.

10. A sequence of three bases on mRNA is called (a) an anticodon; (b) a gene; (c) an amino acid; (d) a codon; (d) a ribosome.

☞ Terminology Exercises

WORD PART	MEANING	WORD PART	MEANING
ana-	apart	-phag-	to eat, devour
cyt-	cell	-phil-	to love, have affinity for
-elle	little, small	-phob-	hate, dislike
extra-	outside, beyond	pino-	to drink
hyper-	excessive, above	-plasm-	matter
hypo-	beneath, below	-reti-	network, lattice
-ic-	pertaining to	-som-	body
intra-	within, inside	ton-	solute strength
iso-	equal, same	-ul-, -ule	small, tiny
-osis	condition of	-um	presence of

Use word parts given above or in previous chapters to form words that have the following definitions.

________________________________ Condition of cell eating

________________________________ Having the same solute strength

________________________________ Condition of attracting water

________________________________ Matter of the cell

________________________________ Within the cell

Using the definitions of word parts given above or in previous chapters, define the following words.

Pinocytosis ________________________________

Somatic ________________________________

Cytology ________________________________

Hydrophobic ________________________________

Endoplasmic ________________________________

Match each of the following definitions with the correct word.

________	Presence of a tiny network	A.	anaphase
________	Excessive solute strength	B.	endocytosis
________	Pulling apart phase	C.	hypertonic
________	Little organs	D.	organelles
________	Condition of taking something into the cell	E.	reticulum

☞ Fun and Games

1. Structure for cell locomotion
2. Stage when chromosomes are pulled apart during mitosis
3. Base that pairs with adenine
4. Slender rods of protein that are part of cytoskeleton
5. Attracts water
6. Granules of RNA in cytoplasm
7. Same solute strength
8. Division of the cytoplasm
9. Diffusion of solvent through a membrane

10A. Slender threads of DNA and protein in nucleus

10D. Organelle within the centrosome

11. Stage in which chromosomes are aligned in center
12. Final stage of mitosis

13A. Short hair like structures for movement

13D. Three bases on mRNA

14. Three bases on tRNA
15. Movement of particles from higher to lower concentration
16. Contains RNA within the nucleus

17A. Gel like fluid within a cell

17D. Response to hypertonic solutions

18. Contain digestive enzymes within the cell
19. Nuclear division in somatic cells
20. Apparatus that "packages" secretory material
21. Base that pairs with guanine

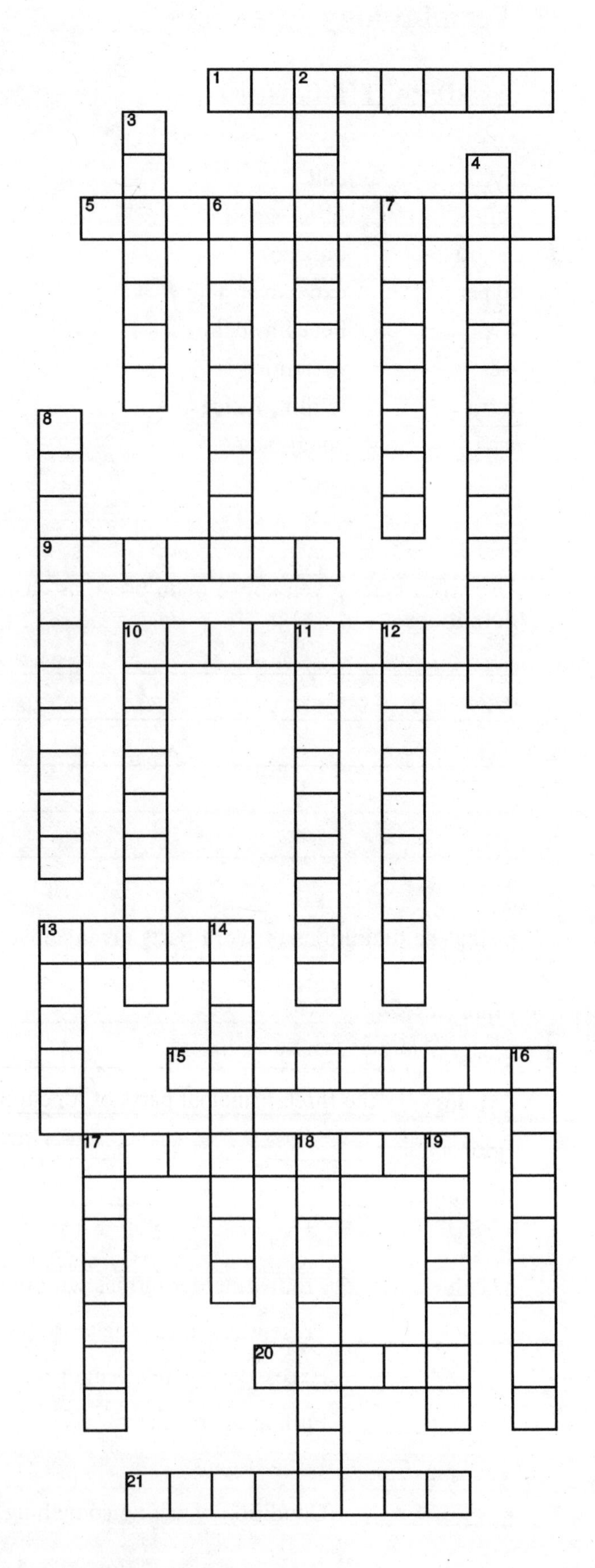

4 Tissues and Membranes

☞ Chapter Outline/Objectives

Body Tissues

1. Explain what is meant by a tissue.
2. Define the term histology.
3. List the four main types of tissues found in the body.
4. Describe epithelial tissues in terms of structure, location, blood supply, and mitotic capabilities.
5. Describe how epithelial tissues are classified according to cell shape and according to the number of layers.
6. Describe each of the following specific epithelial tissues and state at least one location for each: simple squamous, simple cuboidal, simple columnar, pseudostratified columnar, stratified squamous, and transitional.
7. Distinguish between exocrine and endocrine glands.
8. Give an example of a unicellular exocrine gland.
9. Distinguish between simple and compound glands; and between tubular and alveolar glands.
10. Distinguish between merocrine, apocrine, and holocrine glands. Give an example of each.
11. Describe the general characteristics of connective tissues.
12. Distinguish between collagenous fibers and elastic fibers.
13. Name three types of connective tissue cells.
14. Describe the features and location of loose connective tissue, adipose, and dense fibrous connective tissue.
15. Describe the general characteristics of cartilage
16. Distinguish between hyaline cartilage, fibrocartilage, and elastic cartilage. State at least one location for each.
17. Describe the general characteristics and structural features of osseous tissue.
18. Name the intercellular matrix and three types of cells found in blood.
19. Name two types of contractile proteins found in muscle tissue.
20. Distinguish between skeletal muscle, smooth muscle, and cardiac muscle in terms of structure, location, and control.
21. Name two categories of cells in nerve tissue.
22. Identify the three principal parts of a neuron.

Body Membranes

23. Distinguish between epithelial membranes and connective tissue membranes.
24. Describe mucous membranes in terms of structure, secretions, and location.
25. Describe serous membranes in terms of structure, secretions, and location.
26. Identify the two layers found in serous membranes.
27. State the location of the pleura, pericardium, and peritoneum.
28. Name two connective tissue membranes and state a location for each one.
29. Identify the three layers of the meninges.

☞ Learning Exercises

Body Tissues (Objectives 1-22)

1. Write a sentence that defines the term "tissue."

2. The study of tissues is called ______________________________

3. List the four types of tissues found in the body.

   a. c.

   b. d.

4. Complete the description of epithelial tissue by writing the correct words in the blanks.

   Epithelial tissues consist of__________________ cells with________________ intercellular matrix. They have ____________________ surface, have no ____________________ supply, and ____________________ quickly. These tissues __________ the body, ______________ body cavities, and __________ the organs within the body cavities. The cells may be ____________, __________, or ____________________ in shape.

5. Match the following illustrations of epithelial tissues with the correct tissue type and location. Select your answers from the list provided for you.

Type of Epithelium	**Locations**
Simple squamous	Digestive tract
Simple cuboidal	Respiratory passages
Simple columnar with cilia	Urinary bladder
Pseudostratified ciliated columnar	Skin
Stratified squamous	Alveoli of lungs
Transitional	Kidneys

Type______________________

Location______________________

Type________________________

Location______________________

Type________________________

Location______________________

6. Write "endocrine" before each phrase that refers to an endocrine gland and write "exocrine" before each phrase that refers to an exocrine gland.

______________________ Ductless gland

______________________ Goblet cells

______________________ Secretes product onto a surface

______________________ Have ducts

______________________ Secrete product into blood

7. Match each of the following phrases with the term it best describes. There is one answer for each phrase.

_______ Ducts do not have branches

_______ Glandular portion is round

_______ Entire cell is discharged with secretion

_______ No cytoplasm lost with secretion

_______ Portion of cell lost with secretion

A. Compound
B. Simple
C. Acinar
D. Tubular
E. Apocrine
F. Holocrine
G. Merocrine

8. Write the terms that match the following phrases about connective tissue.

______________________	Fibers that are strong and flexible
______________________	Connective tissue cells that are phagocytic
______________________	Type that anchors skin to underlying tissues
______________________	Type of connective tissue that contains fat
______________________	Type that is found in tendons and ligaments
______________________	Covering around cartilage
______________________	Type of cartilage in intervertebral discs
______________________	Most rigid connective tissue
______________________	Structural unit of bone
______________________	Bone cells
______________________	Connective tissue with liquid matrix
______________________	Another name for red blood cells

9. Answer the following questions about the accompanying diagram.

What type of connective tissue is illustrated in this diagram?

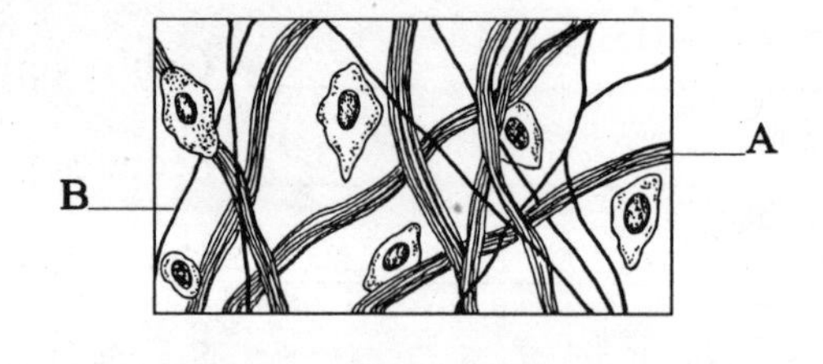

What type of fiber is represented by letter A?

What type of fiber is represented by letter B?

10. Answer the following questions about the accompanying diagram.

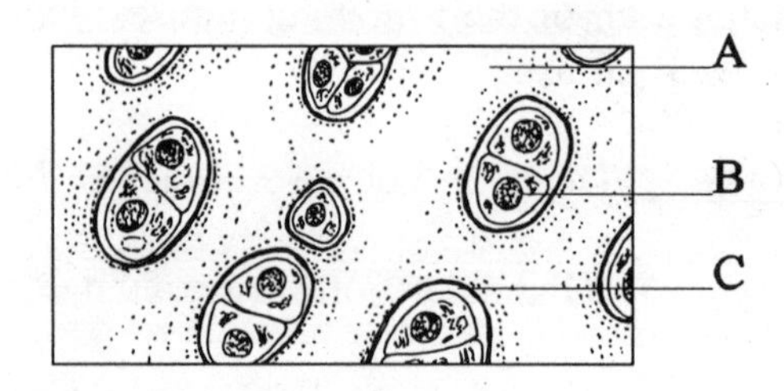

What type of connective tissue is illustrated in this diagram?

What type of protein is found in the intercellular matrix A?

What is the name of the cell represented by letter B?

What is the space represented by letter C?

11. Answer the following questions about the accompanying diagram.

    What type of connective tissue is illustrated in this diagram?

    What term is used for the concentric rings of matrix represented by letter A?

    What is the central structure represented by letter B?

    What term is used for the tiny hairlike structures represented by letter C?

12. Answer the following questions about the accompanying diagram.

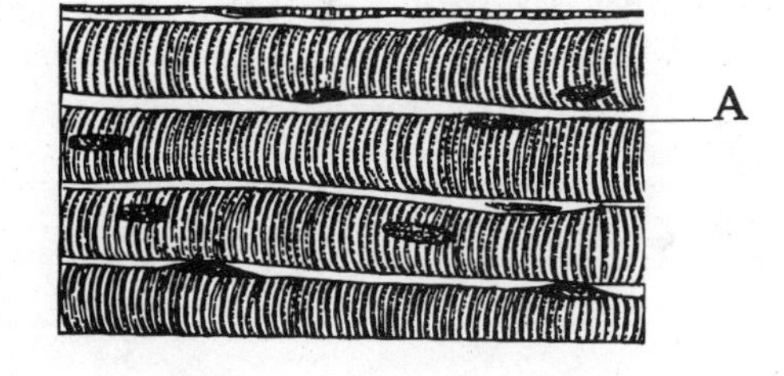

    What type of muscle tissue is illustrated in this diagram?

    What term is used for the cell membrane represented by letter A?

    What two types of contractile proteins are found in this tissue?

13. Answer the following questions about the accompanying diagram.

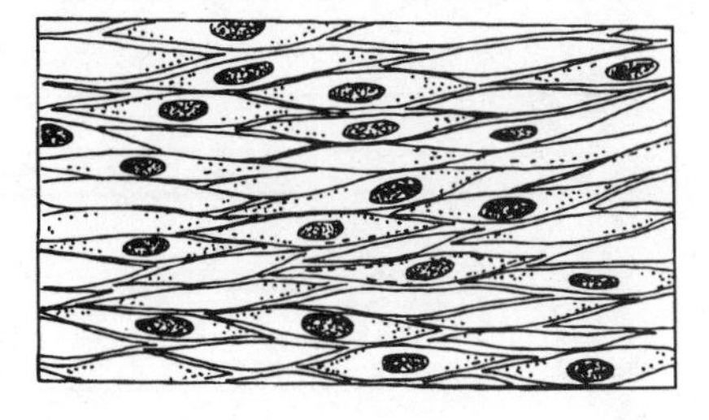

    What type of muscle tissue is illustrated in this diagram?

    Give two locations for this type of muscle tissue.

14. Answer the following questions about the accompanying diagram.

What type of muscle tissue is illustrated in this diagram?

What is the term for the dark band represented by letter B?

Where is this type of muscle located?

B

15. Write the terms that match the following phrases about nervous tissue.

_______________ Cells that transmit impulses

_______________ Processes taking impulses to cell body

_______________ Processes taking impulses away from cell body

_______________ Supporting cells of nerve tissue

Body Membranes (Objectives 23-29)

1. Write the terms that match the following phrases about body membranes.

_______________ Membranes composed only of connective tissue

_______________ Line cavities that open to the outside

_______________ Type of membrane lining digestive tract

_______________ Membranes with two layers

_______________ Layer that lines a cavity wall

_______________ Layer that covers organs within a cavity

_______________ Serous membrane associated with the lungs

_______________ Serous membrane associated with the heart

_______________ Serous membrane of abdominopelvic cavity

_______________ Connective tissue membrane of movable joints

_______________ Connective tissue membranes around the brain

_______________ Tough, outermost layer of meninges

_______________ Middle layer of the meninges

_______________ Innermost layer of the meninges

☞ Chapter Self-Quiz

1. Match each description or component with the correct tissue type.

_______	Closely packed cells	A. Epithelium
_______	Macrophages and fibroblasts	B. Connective
_______	Tendons and ligaments	C. Muscle
_______	Simple and stratified	D. Nerve
_______	Axons and dendrites	
_______	Chondrocytes and osteocytes	
_______	Specialized for contraction	
_______	Blood and bones	
_______	Has a free surface	
_______	Actin and myosin	

2. Match each of the following locations with the correct tissue type. Some responses may be used more than once, others may not be used at all.

_______	Outer layer of skin	A. Adipose
_______	Kidney tubules	B. Dense fibrous connective tissue
_______	Costal cartilage	C. Elastic cartilage
_______	Attaches skin to muscles	D. Fibrocartilage
_______	Lining of the stomach	E. Hyaline cartilage
_______	Most of the fetal skeleton	F. Loose connective tissue
_______	Alveoli of the lungs, for diffusion	G. Simple columnar epithelium
_______	Tendons	H. Simple cuboidal epithelium
_______	Lining of the urinary bladder	I. Simple squamous epithelium
_______	Intervertebral discs	J. Smooth muscle
		K. Stratified squamous epithelium
		L. Transitional epithelium

3. Ductless glands, which secrete their product directly into the blood, are called

_______________________________ glands.

4. Goblet cells are an example of (a) endocrine glands, (b) merocrine glands, (c) apocrine glands, (d) unicellular glands.

5. Ligaments have an abundance of (a) blood vessels, (b) mast cells, (c) elastic fibers, (d) adipose cells, (e) collagenous fibers.

6. Which of the following is found in both osseous tissue and cartilage? (a) canaliculi, (b) lamellae, (c) lacunae, (d) Haversian canals

7. Another name for a platelet is _______________________________.

8. Involuntary control and cross-triations are characteristic of (a) skeletal muscle, (b) smooth muscle, (c) visceral muscle, (d) cardiac muscle.

9. The processes on neurons that carry impulses away from the cell body are called

_______________________________.

10. Membranes that line body cavities that open to the exterior are (a) mucous membranes, (b) synovial membranes, (c) serous membranes, (d) meninges.

☞ Terminology Exercises

WORD PART	MEANING
a-	without, lacking
aden-	gland
adip-	fat
-blast	to form, sprout
chrondr-	cartilage
erythr-	red
fibro-	fiber
glia-	glue
hist-	tissue
leuk-	white
macro-	large
multi-	many
neur-	nerve
-oma	tumor, swelling, mass
oss-	bone
pseudo-	false
squam-	flattened, scale
strat-	layer
thromb-	clot
vas-	vessel

Use word parts given above or in previous chapters to form words that have the following definitions.

____________________________ Cartilage cell

____________________________ Tumor of a gland

____________________________ Cell that forms clots

____________________________ Large phagocytic cell

____________________________ Nerve glue

Using the definitions of word parts given above or in previous chapters, define the following words.

Avascular ____________________________

Fibroblast ____________________________

Erythrocyte ____________________________

Stratified ____________________________

Squamous ____________________________

Match each of the following definitions with the correct word.

__________ Appears to have layers but does not

__________ Tumor of epithelial tissue

__________ Fat tissue

__________ Osseous tissue

__________ Tumor of neuroglia cells

A. adipose
B. bone
C. carcinoma
D. glioma
E. pseudostratified

☞ Fun and Games

Each of the answers in this puzzle is a term from this chapter on tissues and membranes. Fill in the answers to the clues by using syllables from the list that is provided. The number of syllables in each word is indicated by the number in parentheses after the clue. The number of letters in each word is indicated by the number of spaces provided. All syllables in the list are to be used and no syllable is used more than once unless it is duplicated in the list.

A	CYTE	LAGE	O	RON
A	E	LAR	PHAGE	RYTH
AD	EP	LI	PLEU	SQUA
AL	FI	MAC	POSE	SYN
AL	GES	MEN	RA	TA
BRO	GLI	MOUS	RE	THE
CAR	HIS	MINE	RO	TI
CER	I	NEU	RO	UM
CHOND	I	NEU	RO	VI
CYTE	IN	O	RO	VIS

Tissue filled with fat (3) _ _ _ _ _ _ _

Conducts impulses (2) _ _ _ _ _ _

A serous membrane (2) _ _ _ _ _ _

Cartilage cell (3) _ _ _ _ _ _ _ _ _ _

Tissue that forms coverings (5) _ _ _ _ _ _ _ _ _ _

Membranes around the brain (3) _ _ _ _ _ _ _ _

Large phagocytic cell (3) _ _ _ _ _ _ _ _ _ _

Substance in mast cells (3) _ _ _ _ _ _ _ _ _

Red blood cell (4) _ _ _ _ _ _ _ _ _ _ _

Loose connective tissue (4) _ _ _ _ _ _ _

Smooth muscle (3) _ _ _ _ _ _ _ _

Membrane in movable joints (4) _ _ _ _ _ _ _ _

Flat epithelial cells (2) _ _ _ _ _ _ _ _

Tissue in intervertebral discs (5) _ _ _ _ _ _ _ _ _ _ _ _ _ _

Nerve glue (4) _ _ _ _ _ _ _ _ _

5 Integumentary System

☞ Chapter Outline/Objectives

Structure of the Skin

1. Name the components of the integumentary system.
2. Name the two layers of the skin and a third, supporting layer.
3. Describe the structure of the epidermis of the skin.
4. Describe the structure of the dermis of the skin.
5. Give another name for the subcutaneous layer and describe its structure.

Skin Color

6. Discuss three factors that influence skin color.

Epidermal Derivatives

7. Describe the structure of hair and its relationship to the skin.
8. Describe the structure of nails and their relationship to the skin.
9. Discuss th characteristics and functions of sebaceous glands.
10. Distinguish between two types of sudoriferous glands on the basis of distribution and secretory product.
11. Specify the secretory product and location of ceruminous glands.

Functions of the Skin

12. Discuss four functions of the integumentary system.

☞ Learning Exercises

Structure of the Skin (Objectives 1-5)

1. The integumentary system includes ________________, ________________________,

   __________________________, and ____________________________________.

2. Write the name of the layer or structure that corresponds to each of the descriptions. Select your answers from the following choices.

   Dermis
   Epidermis
   Hypodermis
   Stratum basale
   Stratum corneum
   Stratum germinativum
   Stratum granulosum
   Stratum lucidum
   Stratum spinosum

   ________________________ Referred to as the subcutaneous layer

   ________________________ Specific layer that contains melanocytes

   ________________________ Epidermal layer next to the dermis

   ________________________ Consists of stratum basale and stratum spinosum

_______________ Consists of a papillary layer and a reticular layer

_______________ Layer immediately above the stratum basale

_______________ Referred to as superficial fascia

_______________ Found only in thick skin

_______________ Keratinization begins in this layer

_______________ Consists of dead, completely keratinized, cells

_______________ Layer in which hair, nails, and glands are embedded

_______________ Contains receptors for temperature and touch

_______________ Contains adipose tissue

_______________ Actively mitotic layer

_______________ Has collagen fibers that give strength to the skin

3. Identify the layers and structures of the skin in the following diagram by placing the correct letter in the space provided by the listed labels. Then select different colors for each structure and use them to color the coding circles and the corresponding structures in the illustration.

____ Arrector pili muscle ○

____ Blood vessel ○

____ Epidermis ○

____ Hair bulb ○

____ Sebaceous gland ○

____ Stratum basale ○

____ Stratum corneum ○

____ Sweat gland ○

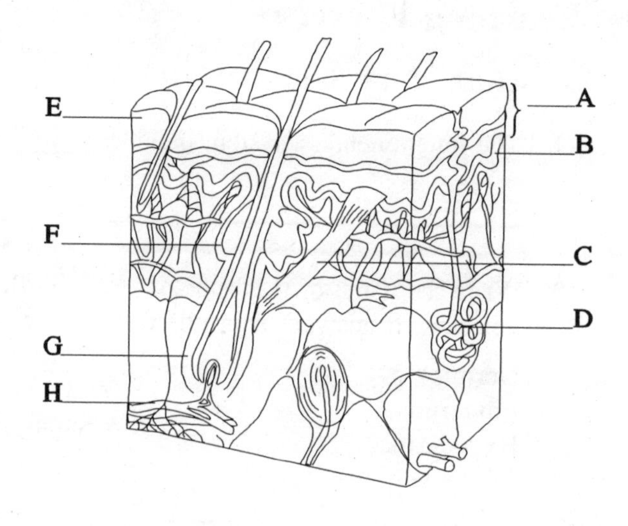

Skin Color (Objective 6)

1. What is the name of the dark pigment that is primarily responsible for skin color?

2. What is the name of the yellow pigment found in the skin?

3. What accounts for the pink color of the skin?

4. Explain why a "tan" is temporary.

Epidermal Derivatives (Objectives 7-11)

Write the terms that fit the following descriptive phrases about epidermal derivatives.

_______________ Tubular sheath surrounding the hair root

_______________ Smooth muscle associated with hair

_______________ Epidermal layer that produces hair and nails

_______________ Central core of a hair

_______________ Crescent-shaped area over the nail matrix

_______________ Type of glands generally associated with hair

_______________ Glands that produce earwax

_____________ Sweat glands that function in temperature regulation

_____________ Another name for sweat glands

_____________ Enlarged region of hair follicle embedded in dermis

_____________ Another name for nail cuticle

_____________ Large sweat glands in the axilla

Functions of the Skin (Objective 12)

1. Four types of functions of the integument are _____________, _____________, _____________, and _____________.
2. The protein that is a waterproofing agent in the skin is _____________.
3. Bacterial growth on the skin is inhibited by _____________.
4. Tissues under the skin are protected from ultraviolet light by _____________.
5. Sense receptors for heat, cold, pain, touch, and pressure are located in the _____________ of the skin.
6. Describe two mechanisms by which the skin functions in temperature regulation.

   (a) Blood vessels

   (b) Sweat glands

7. How does the skin function in the synthesis of vitamin D?

☞ Chapter Self-Quiz

1. Match each of the following descriptions with the correct layer of the skin.

_____	Outermost layer	A. Dermis
_____	Responsible for fingerprints	B. Epidermis
_____	Has five distinct layers in thick skin	C. Hypodermis
_____	Contains adipose	
_____	Has cells responsible for skin color	
_____	Stratified squamous epithelium	
_____	Hair and nails are derived from this layer	
_____	Has sense receptors and hair embedded in it	

2. The central core of hair is the
   (a) medulla
   (b) cortex
   (c) shaft
   (d) follicle
   (e) cuticle.

3. Which of the following is <u>not</u> associated with hair?
   (a) arrector pili
   (b) sebaceous glands
   (c) stratum basale
   (d) apocrine sweat glands
   (e) merocrine sweat glands

4. Which one of the following statements is <u>not</u> true about functions of the skin?
   (a) exposure to ultraviolet light usually increases the activity of melanocytes
   (b) secretions of sweat glands help protect against fluid loss
   (c) when body temperatures increase, the dermal capillaries dilate
   (d) vitamin D is produced in the skin in response to ultraviolet light

5. (True or False) All people have approximately the same number of melanocytes.

☞ Review for Chapters 1-5

How many of these key terms from Chapters 1-5 do you remember? Can you define all of them?

Acid
Active transport
Anatomical position
Anatomy
Arrector pili
Atom
Base
Buffer
Carbohydrate
Ceruminous gland
Chondrocyte
Collagenous fibers
Compound
Covalent bond
Cytokinesis
Dermis
Differentiation
Diffusion
Elastic fibers
Element
Epidermis
Fibroblast
Histology
Homeostasis
Ionic bond
Keratinization
Lipid
Macrophage
Mast cell
Meiosis
Melanin
Metabolism
Mitosis
Molecule
Negative feedback
Neuroglia
Neuron
Osmosis
Osteocyte
Passive transport
Phagocytosis
Physiology
Pinocytosis
Protein
Sebaceous gland
Solute
Solvent
Subcutaneous layer
Sudoriferous layer
Tissue

Can you define these words from the terminology exercises for Chapters 1-5?

Adenoma
Adipose
Alkaline
Anhidrosis
Avascular
Carbohydrate
Carcinoma
Cardiology
Cerumen
Chondrocyte
Cytology
Dermatology
Dermoplasty
Diphosphate
Disaccharide
Dorsal
Endergonic
Endocytosis
Endoplasmic
Epigastric
Erythrocyte
Exergonic
Fibroblast
Gastritis
Glioma
Glycogenesis
Hydrolysis
Hydrophilic
Hydrophobic
Hypodermis
Hyponychium
Ichthyosis
Intracellular
Isotonic
Lactose
Lipogenesis
Lipolysis
Lucidum
Macrophage
Melanocyte
Melanoma
Neuroglia
Onychectomy
Pachyderma
Pachyonychia
Physiology
Pinocytosis
Polysaccharide
Proximal
Pseudostratified
Reticulum
Rhytidoplasty
Skeletal
Somatic
Squamous
Stratified
Subcutaneous
Thrombocyte
Triphosphate
Visceral

☞ Terminology Exercises

WORD PART	MEANING	WORD PART	MEANING
albin-	white	-lucid-	clear, light
cer-	wax	melan-	black
cutane-	skin	onychi-	nail
derm-	skin	pachy-	thick
-ectomy	surgical excision	-plasty	surgical repair
hidr-	sweat	rhytido-	wrinkles
ichthy-	scaly, dry	seb-	oil
kerat-	hard, horny tissue	sud-	sweat

Use word parts given above or in previous chapters to form words that have the following definitions.

______________________________ Black tumor

______________________________ Below the nail

______________________________ Surgical excision of a nail

______________________________ Below the skin

______________________________ Black pigment cell

Using the definitions of word parts given above or in previous chapters, define the following words.

Pachyderma ______________________________

Hypodermis ______________________________

Dermoplasty ______________________________

Dermatology ______________________________

Anhidrosis ______________________________

Match each of the following definitions with the correct word.

________	Clear layer of the skin	A. cerumen
________	Plastic surgery for removal of wrinkles	B. ichthyosis
________	Condition of dry, scaly skin	C. lucidum
________	Earwax	D. pachyonychia
________	Condition of thick nails	E. rhytidoplasty

☞ Fun and Games

Complete this crossword puzzle by filling in the answers to the clues. Clues for words from this chapter are highlighted in bold print.

ACROSS

1. **Dermal layer**
7. **Dead skin layer**
12. Honey or bumble
13. Edge
14. Midwest st.
15. Ages
17. **Yellow pigment**
19. Hunt
21. Allied health career, abb.
22. To travel
23. Opposite of down
24. **Skin layer**
28. Border
31. Grad. class
32. Top ones
34. News source
36. Su'mat, Latin abb.
37. Golf clubs
40. Pointed instrument
42. Large
43. Salt
45. **Pigment cell**
46. You and me
47. Tell's target
49. Prefix for within
50. Ruby or emerald
53. Tablecloths
54. Part of Ner. Sys.
56. 33 1/3 record
57. Chicago transport
58. Astatine symbol
59. Nonrigid airship
62. Inten. care units
64. Scat
65. Actinium symbol
67. Lair
68. 24 hours
70. **Below the skin**
75. Oper. rm.
76. Hoover or Cooley
78. Private teacher
79. See 56 across
80. **Subcutaneous**
83. Denoting presence
84. Repast
86. Illinium symbol
87. Past part. of lie
89. Second cerv. vert.
90. **Narrow lines or bands**
92. Most sick
94. **Integument**
95. Brit. Ortho. Assoc.
96. Fargo st.
97. Univ. of Edinburg
98. Choose
99. **Clear layer**
101. Prompt, hint
102. Woman, Fr.
104. Call for help
105. One at random
106. **Surrounds muscles and organs**
107. Senescent

DOWN

1. Raise
2. Spooky
3. **Skin**
4. Large vessel or vase
5. Falsehood
6. Before noon
7. Female students
8. Prefix for again
9. Agrees with standard
10. See 49 across
11. 50 states
12. Opposite of AD
14. Med. Examiner
16. Male parent
18. Chances, probability
19. **Epidermal layer**
20. **Epidermal protein**
25. Insp. cap.
26. **Oil**
27. **Epidermal layer**
29. River in Egypt
30. **Epidermal layer**
33. Suffix denoting condition
35. **Upper dermal layer**
38. Every night, L., abb.
39. Scandium symbol
41. You and I
44. Lic. Prac. Nurse
48. Escape
51. **Upon the skin**
52. Hertz, abb.
55. **Perspiration**
56. **Moon-shaped nail region**
59. Bill of Health, abb.
60. Promissory note
61. Pacific, abb.
63. Head of a company
66. **Pertaining to skin**
68. Capital of Qatar; to laugh
69. Forearm bones
71. **Bottom layer of epidermis**
72. Sound of disapproval
73. Negative
74. Braced
76. Bring to pass
77. **Dark pigment**
81. Corolla of a flower
82. Hip bone
85. Dog native to Greenland
88. Chicago or Rockford st.
89. Comm. co.
90. Spheno-occipital junction
91. Small whirlpool
93. Research instrument, abb.
95. Public transport vehicle
100. Calcium symbol
101. Cancer, abbrev.
102. Fatty acid
103. Metric unit, abb.

6 Skeletal System

☞ Chapter Outline/Objectives

Overview of the Skeletal System

1. Discuss five functions of the skeletal system.
2. Distinguish between compact and spongy bone on the basis of structural features.
3. Classify bones according to size and shape.
4. Identify the general features of a long bone.
5. Define osteogenesis.
6. Identify three types of cells involved in bone formation and remodeling.
7. Distinguish between intramembranous and endochondral ossification.
8. Describe the processes by which bones increase in length and in diameter.
9. Distinguish between the axial and appendicular skeletons, and state the number of bones in each.

Bones of the Axial Skeleton

10. Identify the bones of the skull and their important surface markings.
11. Describe and identify the hyoid bone.
12. Describe the curvatures of the vertebral column.
13. Identify the general structural features of vertebrae.
14. Compare cervical, thoracic, lumbar, sacral, and coccygeal vertebrae and state the number of each type.
15. Identify the structural features of the ribs and sternum.
16. Distinguish between true ribs and false ribs.

Bones of the Appendicular Skeleton

17. Identify the features of each bone of the pectoral girdle.
18. Identify the bones of the upper extremity and the major features of each bone.
19. Identify the features of the pelvic girdle.
20. Distinguish between the false pelvis and the true pelvis.
21. Identify the bones of the lower extremity and the major features of each bone.

Articulations

22. Compare the structure and function of three types of joints.

☞ Learning Exercises

Overview of the Skeletal System (Objectives 1-9)

1. List five functions of the skeletal system.

   a.

   b.

   c.

   d.

   e.

2. Place an S before each phrase that pertains to spongy bone and a C before each phrase that pertains to compact bone.

   ____ Closely packed osteons

   ____ Contains red bone marrow

   ____ Trabeculae

   ____ Canaliculi radiate from lacunae to osteonic canal

   ____ Contains irregular spaces

3. Place an L before each phrase that pertains to long bones, an S before each phrase that pertains to short bones, and an F before each phrase that pertains to flat bones.

   ____ Has a diploe of spongy bone

   ____ Vertical dimension longer than the horizontal dimension

   ____ Roughly cube-shaped

   ____ Primarily spongy bone covered with a thin layer of compact bone

   ____ Bones in the wrist and ankle

   ____ Bones in the thigh and arm

   ____ Spongy bone between two layers of compact bone

4. Write the terms that fit the following descriptive phrases about the features of long bone.

____________________	Tubular shaft of a long bone
____________________	Expanded ends of long bone
____________________	Outer covering of a long bone
____________________	Hollow region in the shaft
____________________	Hyaline cartilage that covers the ends of the bone
____________________	Connective tissue membrane that lines the shaft

5. Write the terms that fit the following descriptive phrases about the bone markings.

____________________	Smooth, rounded articular surface
____________________	Opening through a bone
____________________	Smooth shallow depression
____________________	Cavity or hollow space in a bone
____________________	Smooth flat articular surface
____________________	Large blunt projections found on the femur

6. Write the terms that fit the following descriptive phrases about bone development and growth.

____________________	Process of bone formation
____________________	Cells that deposit bone matrix; bone-forming cells
____________________	Mature bone cells located in lacunae
____________________	Cells that destroy bone by removing bone matrix
____________________	Type of ossification in most flat bones of the skull
____________________	Bone development in which hyaline cartilage models are replaced by bone
____________________	Region where long bones increase in length
____________________	Cells in the periosteum that are responsible for increase in bone diameter
____________________	Cells that hollow out the medullary cavity
____________________	Type of ossification in most long bones of the body

7. How many named bones are in the complete skeleton? ____________

   How many named bones are in the axial skeleton?__________

   How many named bones are in the appendicular skeleton?__________

Bones of the Axial Skeleton (Objectives 10-16)

1. Name the bones that form the cranium.

2. Name the bones that form the face.

3. For each of the following, name the bone that has the given feature.

____________	Foramen magnum	____________	Optic foramen
____________	Auditory meatus	____________	Cribriform plate
____________	Supraorbital foramen	____________	Sella turcica
____________	Mastoid process	____________	Ramus

4. What is the name of the U-shaped bone in the neck?

5. Write the terms that fit the following descriptive phrases about the axial skeleton.

______________________ First cervical vertebra

______________________ Vertebrae that articulate with ribs

______________________ Weight-bearing portion of a vertebra

______________________ Superior portion of the breastbone

______________________ Second cervical vertebra

______________________ Type of vertebrae in the neck

______________________ Vertebrae with heavy bodies and blunt processes

______________________ Another name for the breastbone

______________________ Cartilaginous pads between the vertebrae

6. State the number for each of the following.

____________	True ribs	____________	Lumbar vertebrae
____________	Cervical vertebrae	____________	Vertebrosternal ribs
____________	False ribs	____________	Vertebral ribs
____________	Thoracic vertebrae	____________	Vertebrochondral ribs

7. Identify the bones and selected features of the skull in the following diagram by placing the correct letter in the space provided by the listed labels. Then select a different color for each bone and use it to color the coding circles and the corresponding bone in the illustration.

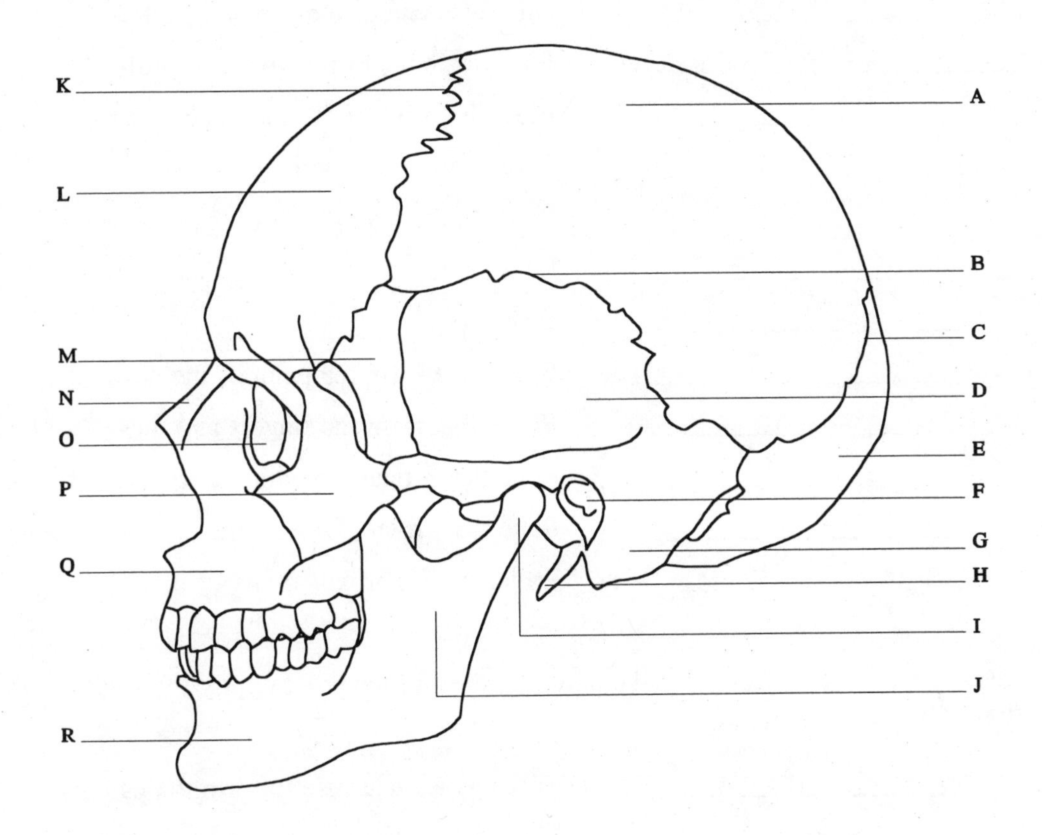

_____	Mandible ○	_____	Parietal ○
_____	Frontal ○	_____	Styloid process
_____	Temporal ○	_____	Auditory meatus
_____	Nasal ○	_____	Mastoid process
_____	Maxilla ○	_____	Mandibular condyle
_____	Occipital ○	_____	Ramus
_____	Zygomatic ○	_____	Coronal suture
_____	Sphenoid ○	_____	Squamosal suture
_____	Lacrimal ○	_____	Lambdoidal suture

Bones of the Appendicular Skeleton (Objectives 17-21)

1. Identify the bone or feature of the pectoral girdle and upper extremity that best fits each description given below.

________________ Anterior bone of the pectoral girdle

________________ Posterior bone of the pectoral girdle

________________ Bone that has an acromion and spine

________________ Large bone in the arm

________________ Bone on the lateral side of the forearm

________________ Bone on the medial side of the forearm

________________ Wrist bones

________________ Bones that form the palm of the hand

________________ Bones that form the fingers and toes

________________ Bone that articulates with the trochlea of humerus

________________ Bone that articulates with the capitulum of humerus

________________ Projection on proximal end of ulna

2. Identify the bone or feature of the pelvic girdle and lower extremity that best fits each description given below.

________________ Three bones that fuse to form the os coxa

____

____

________________ Depression in the os coxa for the femoral head

________________ Portion of femur that articulates with tibia

________________ Bone on the medial side of the leg

________________ Bone on the lateral side of the leg

________________ Tarsal that forms the heel

________________ Tarsal that articulates with the tibia

________________ Bone in tendon anterior to femur and tibia

________________ Bone of axial skeleton that articulates with os coxa

________________ Bone that has greater and lesser trochanters

3. Compare the pelvis in the male and female by completing the following table.

	Male	Female
Pubic arch	____________________	____________________
Pelvic inlet	____________________	____________________
Pelvic cavity	____________________	____________________

4. The portion of the pelvis between the flared wings of the ilium bones is called the ___________ pelvis. The portion inferior to the pelvic brim is the________ pelvis

5. Identify the bones indicated on the anterior view of the skeleton, then color the axial skeleton red and the appendicular skeleton blue.

A. ____________________

B. ____________________

C. ____________________

D. ____________________

E. ____________________

F. ____________________

G. ____________________

H. ____________________

I. ____________________

J. ____________________

K. ____________________

L. ____________________

M. ____________________

N. ____________________

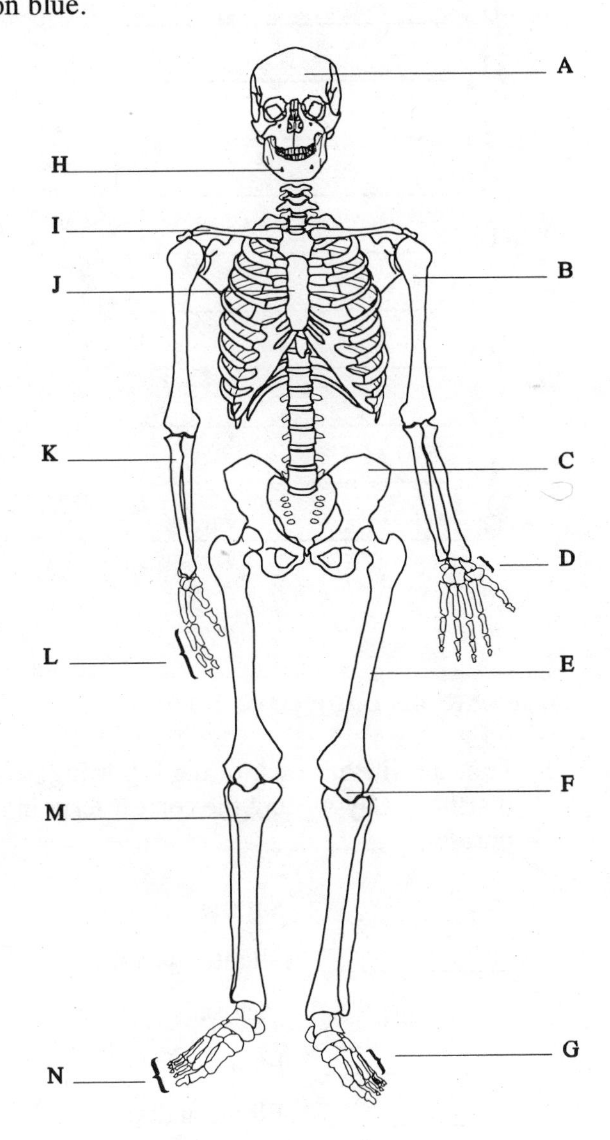

6. Identify the bones indicated on the posterior view of the skeleton, then color the axial skeleton red and the appendicular skeleton blue.

A. ____________________
B. ____________________
C. ____________________
D. ____________________
E. ____________________
F. ____________________
G. ____________________
H. ____________________
I. ____________________
J. ____________________
K. ____________________
L. ____________________
M. ____________________
N. ____________________
O. ____________________
P. ____________________
Q. ____________________

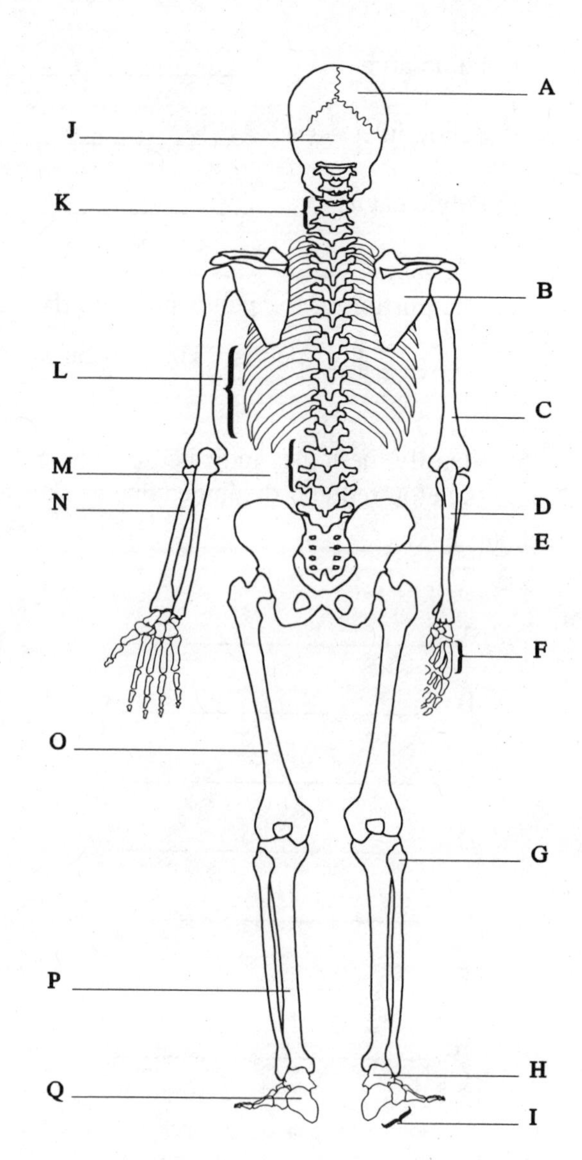

Articulations (Objective 22)

1. Indicate whether each of the following pertains to synarthroses, amphiarthroses, or diarthroses by writing the correct term in the space provided before each descriptive phrase.

____________ Sutures
____________ Slightly movable
____________ Meniscus
____________ Immovable
____________ Elbow and knee
____________ Ball and socket
____________ Ribs to sternum
____________ May have bursae
____________ Joint capsule
____________ Symphysis pubis

☞ Chapter Self-Quiz

1. Match each of the following bones with its correct shape.

______	Carpals	A. Long bone
______	Femur	B. Short bone
______	Metatarsals	C. Flat bone
______	Occipital	D. Irregular bone
______	Phalanges	
______	Sphenoid	
______	Tarsals	
______	Temporal	
______	Tibia	
______	Vertebrae	

2. Match the following descriptions with the correct feature of a long bone.

______	Shaft	A. diaphysis
______	Hollow space in the center	B. endosteum
______	Expanded ends	C. epiphysis
______	Growth region	D. epiphyseal plate
______	Location of yellow marrow	E. medullary cavity
______	Location of red marrow	F. periosteum
______	Outer covering	
______	Lining on inside of shaft	

3. Indicate whether each of the following is part of the axial skeleton or a part of the appendicular skeleton. Use X = axial and P = appendicular.

______	Calcaneus	______	Nasal conchae	______	Scapula
______	Carpals	______	Mandible	______	Sternum
______	Clavicle	______	Occipital	______	Temporal
______	Ethmoid	______	Os coxa	______	Tibia
______	Femur	______	Patella	______	Ulna
______	Humerus	______	Ribs	______	Zygomatic
______	Lumbar vertebrae	______	Sacrum		

4. Which one of the following is <u>not</u> a bone of the cranium? (a) ethmoid, (b) zygomatic, (c) temporal, (d) frontal, (e) sphenoid

5. Which one of the following is <u>not</u> a feature of thoracic vertebrae? (a) centrum or body, (b) long, sharp, spinous process, (c) articular facets for ribs, (d) transverse foramina; (e) vertebral canal

6. If the second item is a part of, or included in, the first, circle the T. If not, circle the F.

T F Occipital bone
Foramen magnum

T F Sphenoid bone
Optic foramen

T F Os coxa
Obturator foramen

T F Ethmoid bone
Inferior nasal conchae

T F Scapula
Acetabulum

T F Humerus
Coronoid process

T F Os coxa
Glenoid fossa

T F Ulna
Styloid process

7. Match the joints, descriptions, or features on the left with the appropriate term on the right.

______ Ball and socket joints

______ Bursae

______ Hinge joints

______ Immovable joints

______ Intervertebral joints

______ Sutures

______ Symphysis pubis

______ Synovial membrane

A. Amphiarthrosis

B. Diarthrosis

C. Synarthrosis

☞ Terminology Exercises

WORD PART	MEANING
acetabul-	little cup
appendicul-	little attachment
artic-	joint
arthr-	joint
-blast	to form, sprout
burs-	pouch
-clast-	to break
corac-	beak
cribr-	sieve
crist-	crest, ridge

WORD PART	MEANING
ethm-	sieve
-fic-	make
kyph-	hump
odont-	tooth
-oid	like, resembling
oste-, oss-	bone
-poie-	making
syn-	together
sphen-	wedge
-tion	act or process of

Use word parts given above or in previous chapters to form words that have the following definitions.

______________________________ Process of making bone

______________________________ Presence of a little cup

______________________________ Resembling a beak

______________________________ Condition of having a hump

______________________________ Break down bone

Using the definitions of word parts given above or in previous chapters, define the following words.

Ethmoid ______________________________

Odontoid ______________________________

Bursa ______________________________

Osteoblast ______________________________

Sphenoid ______________________________

Match each of the following definitions with the correct word.

__________ Condition of making blood — A. Appendicular

__________ Like a sieve — B. Articular

__________ Presence of little attachments — C. Cribriform

__________ Condition of a "together" joint — D. Hemopoiesis

__________ Pertaining to a little joint — E. Synarthrosis

☞ Fun and Games

The names of 50 bones and bone features are hidden in the word search puzzle below. They may read forwards, backwards, horizontally, vertically, or diagonally, but always in a straight line. See how many you can find.

```
S F V H D M A S T O I D Q S T E M P O R A L
S U P E W X N A S A L W F M U I H C S I P S
X N T Y R Z M S I F R W O V B I L I U M D M
C R C A R T I B N I T S R E E W D L N G X B
I D L K E A E P F B S L A S R A T A T E M O
T T A P L M L B E U I V M L O H I U R R G N
A T V H H A Y L R L L G E S S B P U B I S R
M G I S C A M R I A M A N D I B L E D L M U
O U C J T A L I O X E G M T T X V H A C Y Y
G H L H C T R A R T A P A R Y F A P O E P H
Y U E U O A L P N C I M G O T K R C C A W V
Z M M M B U L S A G A D N C M A C F L M Q P
O E L U P A A C S L E L U H C Y Q A U M S A
N R K A I L T U A Q S S M A X N T R H P C L
Q U C O T R F E L N C P T N L I C Y H M O A
E S P A H I B R C C E E S T N A O E Y U N G
H Z A F K S P U O A M U O E S O N W D N D I
H N T V E O D I N N X X S R I O J R C R Y E
G S E W S M J L C A T O E H I F R W E E L X
M O L T A L U S H C M A C D D I O M H T E B
X B L A T E I R A P O K L S B B O N L S X S
A H A X I P H O I D R T G S O V X G C A J E
```

7 Muscular System

☞ Chapter Outline/Objectives

Characteristics and Functions of the Muscular System

1. Compare skeletal muscle tissue with visceral and cardiac muscle tissue.
2. List four characteristics of muscle tissue that relate to its functions.
3. List three functions of skeletal muscle tissue.

Structure of Skeletal Muscle

4. Describe the structure of a skeletal muscle, including its connective tissue coverings.
5. Define the terms origin and insertion as they relate to muscle function.
6. Distinguish between actin and myosin.
7. Identify the bands and lines on myofibers that make up the striations of skeletal muscle.
8. Define the term sarcomere.
9. Explain why skeletal muscle needs an abundant nerve and blood supply.

Contraction of Skeletal Muscle

10. Describe the sequence of events at the neuromuscular junction that provide the stimulus for contraction of a sarcomere.
11. Describe the changes that take place in actin and myosin myofilaments during contraction.
12. Use the terms threshold and subthreshold stimulus to explain what is meant by the all-or-none principle as it pertains to skeletal muscle contraction.
13. Define the terms motor unit, multiple motor unit summation, twitch, tetanus, multiple wave summation, treppe, and muscle tone.
14. Distinguish between isometric and isotonic contractions.
15. Describe how energy is provided for muscle contraction and how oxygen debt occurs.
16. Distinguish between synergists and antagonists.
17. Describe and illustrate flexion, extension, hyperextension, dorsiflexion, plantar flexion, abduction, adduction, rotation, supination, pronation, circumduction, inversion, and eversion.

Skeletal Muscle Groups

18. Describe seven features of a muscle that are frequently used to name the muscle.
19. Locate, identify, and describe the actions of the major muscles of the head and neck.
20. Locate, identify, and describe the actions of the major muscles of the trunk.
21. Locate, identify, and describe the actions of the major muscles of the upper extremity.
22. Locate, identify, and describe the actions of the major muscles of the lower extremity.

☞ Learning Exercises

Characteristics and Functions of the Muscular System (Objectives 1-3)

1. Match the correct type of muscle tissue with each descriptive word or phrase.

____________	Striated and involuntary	A. Skeletal
____________	Spindle-shaped fibers	B. Visceral
____________	Multinucleated and cylindrical	C. Cardiac
____________	Found in blood vessels	
____________	Found in the heart	

2. List four characteristics of muscle tissue that relate to its functions.

   a. ______________________________ c. ______________________________

   b. ______________________________ d. ______________________________

3. Three functions of muscle contraction are:

   a. ______________________________

   b. ______________________________

   c. ______________________________

Structure of Skeletal Muscle (Objectives 4-9)

1. Write the terms that fit the following descriptive phrases in the space provided.

______________________________	More movable attachment of a muscle
______________________________	Covering around an individual muscle fiber
______________________________	Protein in thick myofilaments
______________________________	Bundle of fibers surrounded by perimysium
______________________________	Less movable attachment of a muscle
______________________________	Broad, flat sheet of tendon
______________________________	Cell membrane of a muscle cell
______________________________	Protein in thin myofilaments
______________________________	Unit of muscle between Z lines
______________________________	Type of myofilament in the I band

2. Muscle fibers must be stimulated before they can contract, therefore they have an abundant ______________________________. The blood supply delivers ____________________ and ______________________________ for contraction.

Contraction of Skeletal Muscle (Objectives 10-17)

1. Arrange the following events of muscle contraction in the correct sequence by writing a number 1 before the first event, number 2 before the second event, and so forth until the 8 events have been numbered correctly.

_______ Energized myosin heads attach to actin

_______ Acetylcholine is released into synaptic cleft

_______ Power stroke pulls actin toward center of A band

_______ Calcium reacts with troponin and exposes binding sites on actin

_______ Acetylcholine reacts with receptors on sarcolemma

_______ Nerve impulse reaches axon terminal

_______ Calcium ions are released from the sarcoplasmic reticulum

_______ Impulse travels into the T tubules

2. Write the terms that match the statements in the spaces provided.

_______________________ Minimum stimulus that causes contraction

_______________________ Principle by which muscle fibers contract

_______________________ Stimulus insufficient to cause contraction

_______________________ Single neuron and muscle fibers it stimulates

_______________________ Increases contraction strength in a muscle

_______________________ Sustained contraction due to rapid stimuli

_______________________ Staircase effect due to a series of stimuli

_______________________ Continued state of partial muscle contraction

_______________________ Muscle contraction with constant tension

_______________________ Muscle contraction with changing tension

3. Write the terms that match the statements in the spaces provided.

_______________________ Immediate source of energy for contraction

_______________________ Stored in muscle to regenerate ATP

_______________________ Molecule that stores oxygen in muscle

_______________________ Acid that accumulates with lack of oxygen

_______________________ Products of aerobic breakdown of pyruvic acid

4. The muscle that has a primary role in providing a movement is called the _____________. Muscles that assist this muscle are called _______________________ and muscles that oppose the movement are called _______________________.

5. Match the type of muscle movement with the description.

	Description		Movement
____	Extension of foot to stand on tiptoes	A.	Flexion
____	Movement of arm toward the midline	B.	Extension
____	Straightening the arm at the elbow	C.	Hyperextension
____	Turning the palm of the hand forward	D.	Dorsiflexion
____	Turning the head from side to side	E.	Plantar flexion
____	Moving elbow to put hand on shoulder	F.	Abduction
____	Turning the sole of foot inward	G.	Adduction
____	Spreading the fingers apart	H.	Rotation
____	Tilting head backward	I.	Supination
____	Drawing circles on chalkboard	J.	Pronation
		K.	Circumduction
		L.	Inversion
		M.	Eversion

Skeletal Muscle Groups (Objectives 18-22)

1. Write the type of muscle feature that is indicated by each of the following muscle names.

_______________ Adductor

_______________ Rectus

_______________ Maximus

_______________ Gluteus

_______________ Deltoid

_______________ Biceps

_______________ Brachioradialis

_______________ Pectoralis

2. Write the name of the muscle that matches each action.

___________________________ Raises the eyebrows

___________________________ Closes and puckers the lips

___________________________ Raise the mandible in chewing (2)

____

___________________________ Closes the eye in squinting

___________________________ Flexes neck so the chin drops

___________________________ Primary muscle in inspiration

___________________________ Anterior muscle that adducts and flexes arm

___________________________ Posterior muscle that adducts arm

___________________________ Abducts the arm

___________________________ Extends the forearm

___________________________ Has two heads and flexes forearm

___________________________ Inserts on ulna and flexes forearm

___________________________ Inserts on femur; flexes thigh

________________ Synergist of iliopsoas

________________ Antagonist of iliopsoas

________________ Group that adducts thigh

________________ Group that extends the leg

________________ Muscles that flex the leg (3)

____

____

________________ Two muscles that plantar flex the foot

____

________________ Antagonist of the gastrocnemius

3. Identify the muscles indicated on the following diagrams.

A. ________________

B. ________________

C. ________________

D. ________________

E. ________________

F. ________________

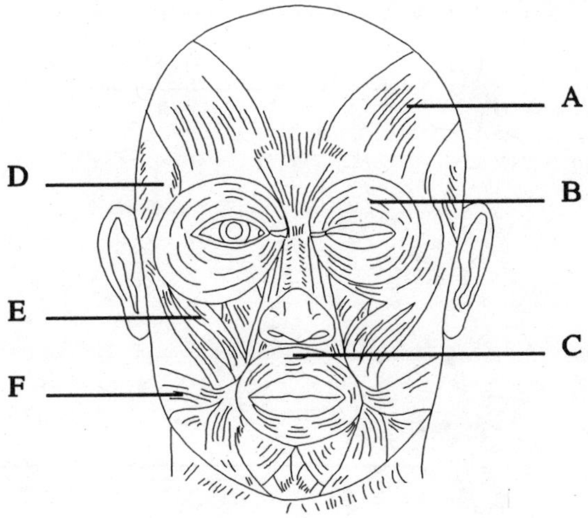

G. ________________

H. ________________

I. ________________

J. ________________

K. ________________

L. ________________

M. ________________

N. ________________

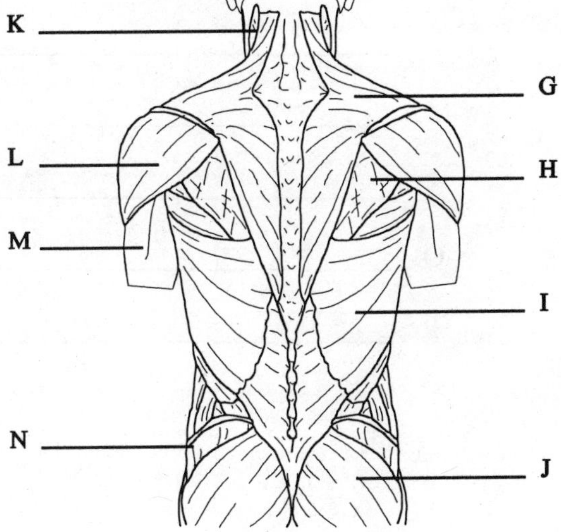

4. Identify the muscles indicated on the following diagrams.

A. ______________________

B. ______________________

C. ______________________

D. ______________________

E. ______________________

F. ______________________

G. ______________________

H. ______________________

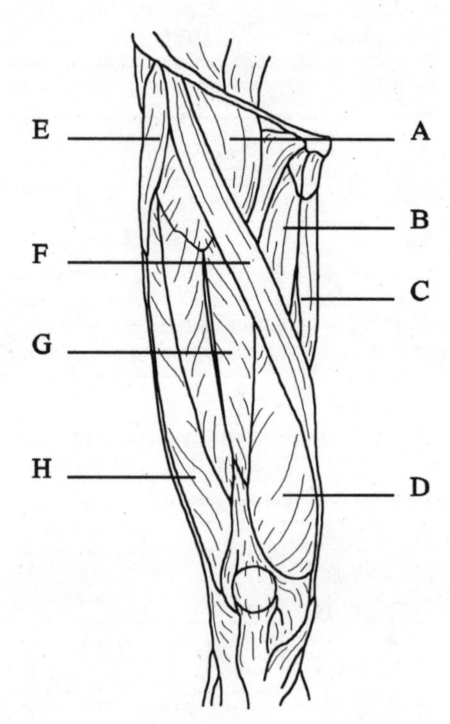

I. ______________________

J. ______________________

K. ______________________

L. ______________________

M. ______________________

N. ______________________

O. ______________________

P. ______________________

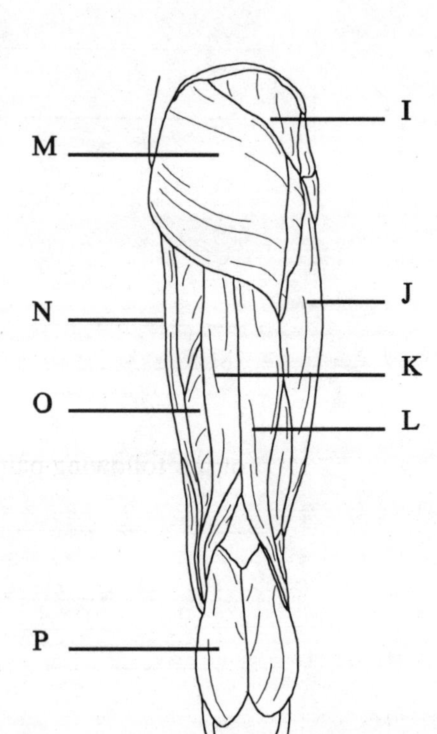

☞ Chapter Self-Quiz

1. Place an X before the characteristics that pertain to skeletal muscle tissue.

   _____ Responds to a stimulus

   _____ Multinucleated

   _____ Centrally located nucleus

   _____ Striations

   _____ Intercalated discs

   _____ Located in the wall of the stomach

2. The connective tissue covering around a fasciculus of skeletal muscle fibers is called ________________________.

3. The fixed, or stable, end of a muscle is called the______________________.

4. The region of a myofibril that has only actin filaments is the (a) A band; (b) I band; (c) H band; (d) M band; (e) Z band.

5. Name the following:

   ________________________ Cell membrane of a muscle cell

   ________________________ Neurotransmitter at neuromuscular junction

   ________________________ Protein in thick filaments

   ________________________ Molecule that accumulates in anaerobic respiration

   ________________________ Storage site for calcium ions

   ________________________ Immediate energy source for contraction

   ________________________ Type of contraction that produces movement

   ________________________ Muscle response to a single stimulus

   ________________________ A type of multiple wave summation

   ________________________ Staircase effect of contraction

6. Which of the following pairs of terms does <u>not</u> represent opposite actions? (a) flexion - extension; (b) rotation - pronation; (c) inversion - eversion; (d) abduction - adduction; (e) dorsiflexion - plantar flexion

7. When you bend the elbow to touch your right shoulder with your right hand, the action at the elbow is (a) abduction; (b) supination; (c) circumduction; (d) flexion; (e) inversion.

8. Match the following descriptions with the appropriate muscle.

A. Adductor longus
B. Biceps brachii
C. Deltoid
D. Gastrocnemius
E. Hamstrings
F. Masseter
G. Pectoralis major
H. Quadriceps femoris
I. Rectus abdominis
J. Sartorius
K. Serratus anterior
L. Tensor fasciae latae
M. Tibialis anterior
N. Trapezius
O. Triceps brachii
P. Zygomaticus

_____ Closes the jaw in mastication
_____ Elevates the corner of the mouth in smiling
_____ Extends the head; holds the head upright
_____ Long, straight muscle of the anterior abdomen
_____ Moves the scapula
_____ Adducts the arm
_____ Antagonist of the latissimus dorsi
_____ Inserts on the radius
_____ Antagonist of the biceps brachii
_____ Located in medial compartment of thigh
_____ Most superficial and lateral muscle in the thigh
_____ Long straplike muscle that flexes thigh and leg
_____ Muscle group that extends the leg at the knee
_____ Located in the posterior compartment of the leg
_____ Muscle group that flexes the leg at the knee
_____ Muscle that allows you to "walk on your heels"

☞ Terminology Exercises

WORD PART	MEANING
a-	without, lacking
act-	motion
bi-	two
cep-	head
delt-	triangle
dia-	through
duct-	movement
flex-	bend
-in	neutral substance
iso-	same, alike

WORD PART	MEANING
lemm-	peel, rind
masset-	chew
metr-	measure
myo-, mys-	muscle
phragm-	fence, partition
sarco-	flesh, muscle
syn-	together
tetan-	stiff
ton-	tone, tension
troph-	nourish, develop

Use word parts given above or in previous chapters to form words that have the following definitions.

_______________________ Neutral substance in motion

_______________________ Partition through a space

_______________________ Same measure

_______________________ Chewer

_______________________ Movement away from

Using the definitions of word parts given above or in previous chapters, define the following words.

Deltoid _______________________

Biceps _______________________

Isotonic _______________________

Flexion _______________________

Sarcolemma _______________________

Match each of the following definitions with the correct word.

______	Neutral substance in muscle	A. Atrophy
______	Work together	B. Myosin
______	Muscle matter	C. Sarcoplasm
______	Without development, wasting	D. Synergist
______	Stiff	E. Tetany

☞ Fun and Games

The object of this puzzle is to accumulate as many points as possible for the words you select as answers for the clues. To do the puzzle, answer each clue with a **single** word and write that word in the space by the clue. Each letter of the alphabet is assigned point values as indicated. Using these point values, add up your score for each answer. Each clue has more than one possible answer and you should try to choose the one that gives the highest point value. Finally, add the ten individual scores to get your total score for the puzzle. For fair play, use single word answers only and avoid answers, such as orbicularis oris and pectoralis major, that contain two words. Try competing with your classmates to see who can get the highest score! Have fun!!

A = 1	B = 2	C = 2	D = 2	E = 1	F = 3	G = 3
H = 3	I = 1	J = 5	K = 4	L = 2	M = 2	N = 1
O = 1	P =3	Q = 5	R = 1	S = 1	T = 1	U = 1
V = 4	W = 4	X = 5	Y = 3	Z = 5		

Clue	Single Word Answer	Points
1. A characteristic of muscle tissue	______	___
2. Connective tissue coverings associated with muscles and muscle fibers	______	___
3. A contractile protein in muscle	______	___
4. Attachment of a muscle	______	___
5. Part, or phase, of a muscle twitch	______	___
6. Descriptive term to depict a particular movement	______	___
7. Word in a muscle name that describes the direction of the fibers	______	___
8. A muscle of facial expression	______	___
9. A muscle that moves the brachium	______	___
10. A muscle in the posterior compartment of the leg	______	___

8 Nervous System

☞ Chapter Outline/Objectives

Functions and Organization of the Nervous System

1. Describe three types of nervous system functions.
2. Outline the organization of the nervous system.

Nerve Tissue

3. Identify the two categories of cells in nerve tissue.
4. Describe the structure of a neuron.
5. Identify the parts of a neuron on a diagram or model.
6. Classify neurons according to their function.
7. Name, identify the location, and state the functions of six types of neuroglia cells.

Nerve Impulses

8. State two functional characteristics of neurons.
9. Describe the resting polarity of the resting cell membrane of a neuron.
10. Describe the sequence of events that lead to an action potential when the cell membrane is stimulated.
11. Explain how the cell membrane is restored to resting conditions after an action potential.
12. Define the terms threshold stimulus and subthreshold stimulus as they pertain to action potentials.
13. Explain how an impulse is conducted along the length of a neuron.
14. Define the terms saltatory conduction, refractory period, and all-or-none principle as they pertain to nerve impulses.
15. Describe the structure of a synapse and how an impulse is conducted from one neuron to another across the synapse.
16. Distinguish between excitatory transmission and inhibitory transmission.
17. Describe three types of circuits in neuronal pools.
18. List the five basic components of a reflex arc.

Central Nervous System

19. Describe the three layers of meninges around the central nervous system.
20. Locate and identify the lobes of the cerebral hemispheres, gray matter, white matter, and basal ganglia.
21. Locate the primary motor, sensory, and association areas of the cerebrum.
22. Describe the location, structure, and functions of the diencephalon.
23. Identify the three regions of the brainstem and describe the functions of each region.
24. Describe the structure and functions of the cerebellum.
25. Trace the flow of cerebrospinal fluid from its origin in the ventricles to its return to the blood.
26. Describe the structure and functions of the spinal cord.

Peripheral Nervous System

27. Use the terms endoneurium, perineurium, epineurium, and fasciculus to describe the structure of a nerve.
28. List the twelve cranial nerves and state the function of each one.
29. Describe the composition of spinal nerves, list the five groups of spinal nerves, and state the number of nerves in each group.
30. Define the term plexus and identify at least one nerve that arises from each of the cervical, brachial, and lumbosacral plexuses.
31. Compare the structural and functional differences between the somatic efferent pathways and the autonomic nervous system.
32. Distinguish between the sympathetic and parasympathetic divisions of the autonomic nervous system in terms of structure, function, and neurotransmitters.

☞ Learning Exercises

Functions of the Nervous System (Objective 1)

1. The activities of the nervous system are grouped into three functional categories. These are ____________________, ________________, and ____________________.

Organization of the Nervous System (Objective 2)

1. The two components of the central nervous system are the ________ and ____________.
2. The two divisions of the peripheral nervous system are the ________, or sensory division, and the __________________, or motor division.
3. The autonomic nervous system is a part of the __________________ division of the peripheral nervous system.

Nerve Tissue (Objectives 3-7)

1. Write the terms that fit the following descriptive phrases about neurone.

______________________ Process that conducts impulses toward the cell body

______________________ Fatty substance around some axons

______________________ Neurons that carry impulses toward the CNS

______________________ Gaps in the fatty substance around axons

______________________ Neurons entirely within the CNS

______________________ Conducts impulses away from the cell body

______________________ Has important role in nerve fiber regeneration

2. Label the following diagram of a neuron.

A. ______________________

B. ______________________

C. ______________________

D. ______________________

E. ______________________

F. ______________________

G. ______________________

H. ______________________

I. ______________________

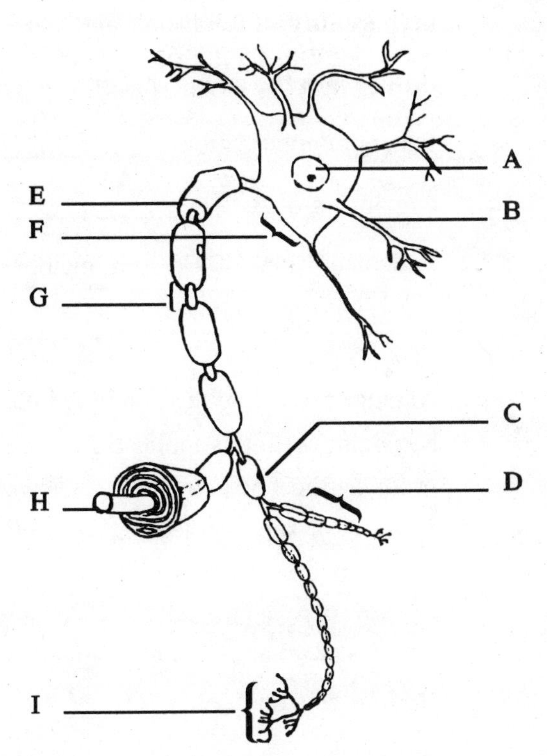

3. Write the name of the neuroglial cell that best fits each descriptive phrase.

______________________ Binds blood vessels to nerves

______________________ Phagocytic

______________________ Produces myelin within the CNS

______________________ Produces myelin within the PNS

______________________ Facilitates circulation of CSF

______________________ Supports cell bodies within ganglia

Nerve Impulses (Objectives 8-18)

1. Two functional characteristics of neurons are ______________________ and ______________________.

2. At rest, the outside of a neuron is ______________ (charge) and has a higher concentration of ______________ ions relative to the inside. A stimulus changes the permeability of the membrane and it depolarizes. During depolarization, __________ ions diffuse into the cell. This creates an ______________ potential, or nerve impulse. During repolarization, ______________ ions diffuse out of the cell. At the conclusion of the ______________, the ______________ pump restores the ionic conditions of a resting membrane. The minimum stimulus required to initiate a nerve impulse is called a ______________ stimulus.

3. Arrange the following numbered elements of a reflex arc in the correct sequence, beginning with the application of a stimulus, by writing the numbers in the appropriate order on the line provided. 1) Axon of interneuron; 2) Axon of motor neuron; 3) Axon of sensory neuron; 4) Cell body of interneuron; 5) Cell body of motor neuron; 6) Cell body of sensory neuron; 7) Dendrite of interneuron; 8) Dendrite of motor neuron; 9) Dendrite of sensory neuron; 10) Effector; 11) Sensory receptor; 12) Synapse between interneuron and motor neuron; 13) Synapse between interneuron and sensory neuron

___ ___ ___ ___ ___ ___ ___ ___ ___ ___ ___ ___ ___

4. Match each of the following terms with the correct descriptive phrase.

A. Convergence
B. Divergence
C. Excitatory
D. Inhibitory
E. Neurotransmitter
F. Reflex arc
G. Refractory period
H. Saltatory
I. Synapse

_______ Rapid conduction from node to node

_______ Region of communication between two neurons

_______ Diffuses across synaptic cleft

_______ Time during which a neuron is recovering from depolarization

_______ Synaptic transmission that initiates an impulse on the postsynaptic membrane

_______ Functional unit of the nervous system

_______ Single neuron synapses with multiple neurons

_______ Synaptic transmission that makes it more difficult to generate an impulse

_______ Several neurons synapse with a single posynaptic neuron

Central Nervous System (Objectives 19-26)

1. Arrange the following spaces and CNS coverings in the correct sequence by placing numbers 1-6 on the lines provided. Use 1 for the **outermost** and 6 for the **innermost**. Color the squares by the meninges red. Color the squares by the spaces blue. Circle the one that is the location of cerebrospinal fluid.

_______ Arachnoid ☐ _______ Epidural ☐ _______ Subarachnoid ☐

_______ Dura mater ☐ _______ Pia mater ☐ _______ Subdural ☐

2. Identify the regions of the brain indicated on the following diagram. Color the somatosensory area red, the somatomotor area blue, visual area green, and the brainstem yellow.

A. _______________

B. _______________

C. _______________

D. _______________

E. _______________

F. _______________

G. _______________

H. _______________

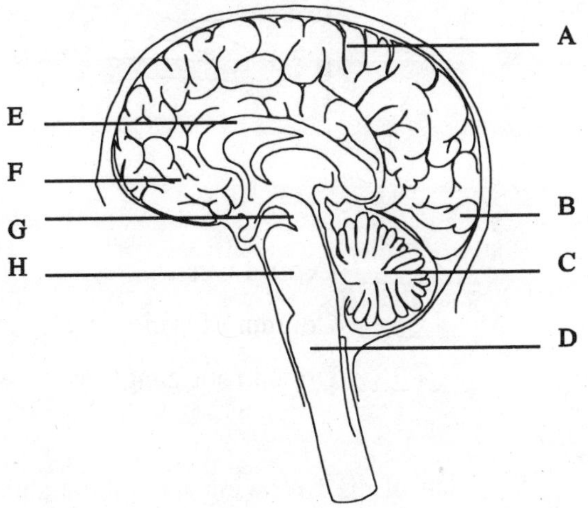

A
E
F
B
G
H
C
D

3. Write the name of the CNS structure that best fits each descriptive phrase.

_______________ Second largest portion of the brain

_______________ Includes the midbrain, pons, and medulla oblongata

_______________ Includes the thalamus and hypothalamus

_______________ Extends from the foramen magnum to the pons

_______________ Diencephalon region regulating visceral activities

_______________ Contains the apneustic and pneumotaxic centers

_______________ Superior portion of the brainstem

_______________ Coordinates skeletal muscles, posture, and balance

_______________ Contains cardiac, vasomotor, respiratory centers

_______________ Middle portion of the brainstem

4. Arrange the following numbered items in the appropriate sequence to represent the correct order of flow for cerebrospinal fluid. Write the numbers in the appropriate order on the line provided.

   1) Openings in the fourth ventricle; 2) Interventricular foramen; 3) Subarachnoid space; 4) Fourth ventricle; 5) Superior sagittal sinus; 6) Third ventricle; 7) Choroid plexus of lateral ventricle; 8) Arachnoid granulations; 9) Cerebral aqueduct

   ____ ____ ____ ____ ____ ____ ____ ____ ____

5. Identify the parts of a spinal cord and spinal nerve that are indicated by leader lines on the following diagram by writing the correct letter in the space provided before each term.

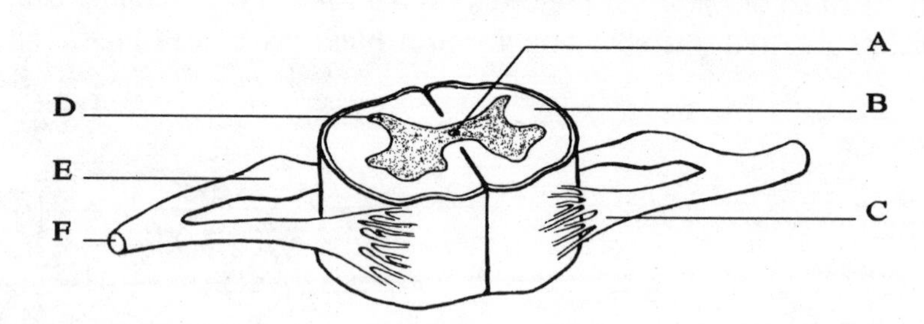

_______	Central canal	______	Dorsal horn
_______	Column of white matter	______	Spinal nerve
_______	Dorsal root ganglion	______	Ventral root

6. Each of the following statements about the spinal cord is false. Rewrite the statements to make them true.

   The spinal cord consists of a central core of white matter surrounded by gray matter.

   Two enlargements of the spinal cord are in the thoracic and lumbar regions.

   The dorsal, lateral, and ventral horns contain bundles of nerve fibers, called nerve tracts.

   Ascending tracts carry motor impulses to the brain.

   Corticospinal tracts are ascending tracts that begin in the cerebral cortex.

Peripheral Nervous System (Objectives 27-32)

1. Within a nerve, each individual nerve fiber has a connective tissue covering called ________________________. The nerve itself is covered by connective tissue called ______________________________.

2. Name the following cranial nerves.

   Three nerves that are sensory only.

   Nerve to the muscles of facial expression.

   Three nerves that function in eye movement.

   Nerve that is likely to be involved if you have a toothache in the lower jaw.

   Nerve to the muscles of mastication.

   Nerve that allows you to nod your head.

3. Write the term or number that best fits each descriptive phrase.

   ________________________ Number of spinal nerves

   ________________________ Spinal nerve root that contains only sensory fibers

   ________________________ Complex network of nerve fibers

   ________________________ Number of cervical nerves

   ________________________ Spinal nerve root that contains only motor fibers

   ________________________ Nerve plexus that supplies nerves to the arm

   ________________________ Nerve plexus that gives rise to the phrenic nerve

   ________________________ Nerve plexus that gives rise to the sciatic nerve

4. Two types of efferent pathways are illustrated below. (A) Color the arrow(s) of the somatic motor pathway red and the autonomic pathway blue; (B) Identify the arrows that represent preganglionic and postganglionic fibers by writing the words in the arrow space; (C) Put an asterisk (*) by the effector that represents smooth muscle, cardiac muscle, and glands.

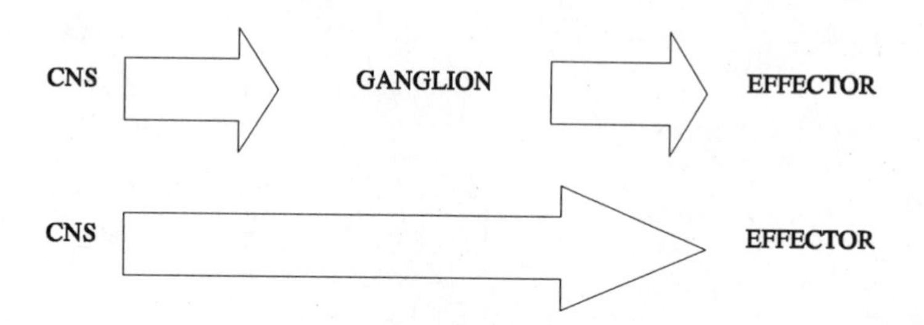

5. The following table indicates some features of the autonomic nervous system. Use a check (✓) to show which division of the autonomic nervous system is involved with each feature. Both divisions may be involved in some features and both columns should be checked. S = Sympathetic Division P = Parasympathetic division

Feature	S	P
Arises from thoracic and lumbar regions of the spinal cord		
Has terminal ganglia		
Also called the craniosacral division		
Has short preganglionic fibers		
Also called "fight-or-flight"		
Cholinergic preganglionic fibers		
Adrenergic postganglionic fibers		
Also called "rest-and-repose" system		
Dilates pupils of the eyes		
Has short postganglionic fibers		
Increases heart rate		
Cholinergic postganglionic fibers		
Dilates blood vessels to skeletal muscles		
Increases digestive enzymes		
Constricts the bronchi		

☞ Chapter Self-Quiz

1. Which of the following are components of the central nervous system? (a) brain and cranial nerves; (b) cranial nerves and spinal nerves; (c) spinal cord and spinal nerves; (d) brain and spinal cord; (e) sensory division and motor division

2. Name each of the following:

   ________________ Fatty covering around a neuron process

   ________________ Afferent neuron process

   ________________ Neuroglia cell that is phagocytic

   ________________ Cell that produces myelin in the CNS

   ________________ Neuron process that may have a sense receptor at its distal end

3. Number the following events in the sequence in which they occur.

   __________ Membrane becomes permeable to potassium

   __________ Threshold stimulus is applied

   __________ Sodium/potassium pump restores resting conditions

   __________ Inside of membrane is positive to the outside

   __________ Membrane becomes permeable to sodium

   __________ Membrane is polarized with potassium outside

4. Which of the following is <u>not</u> true about synaptic transmission? (a) neurotransmitters react with receptor sites on the postsynaptic membrane; (b) impulses "jump" across the synapse by saltatory conduction; (c) inhibitory neurotransmitters hyperpolarize the cell membrane; (d) excitatory neurotransmitters depolarize the cell membrane

5. Name the following:

   ________________ Outermost layer of meninges

   ________________ Structure that produces cerebrospinal fluid

   ________________ Ventricle that is midline in the region of the diencephalon

   ________________ Connective tissue covering an individual nerve fiber

   ________________ Band of white fibers that connects the two cerebral hemispheres

6. Which one of the following is <u>not</u> a correctly matched pair of terms? (a) cervical plexus - phrenic nerve; (b) lumbosacral plexus - median nerve; (c) brachial plexus - radial nerve; (d) lumbosacral plexus - sciatic nerve; (e) brachial plexus - axillary nerve

7. Match each of the following with the region of the brain in which it is found.

_______ Arbor vitae
_______ Pons
_______ Insula
_______ Primary motor area
_______ Basal ganglia
_______ Thalamus
_______ Vermis
_______ Midbrain
_______ Lateral ventricles
_______ Cardiac center

A. Cerebrum
B. Cerebellum
C. Diencephalon
D. Brainstem

8. Which cranial nerve is used for each of the following?

_______________ Transmits sensory impulses from an aching tooth
_______________ Transmits visual impulses from the retina of the eye
_______________ Transmits motor impulses to muscles for smiling
_______________ Transmits motor impulses that affect heart rate
_______________ Transmits motor impulses to the tongue

9. Indicate whether each of the following pertains to the (S) sympathetic, (P) parasympathetic, or (B) both divisions of the autonomic nervous system.

__________ Terminal ganglia
__________ Two neurons in the pathway
__________ Cholinergic fibers
__________ Short preganglionic fibers
__________ Adrenergic fibers
__________ Conserves energy
__________ Craniosacral division
__________ Widespread and long-lasting effect
__________ Shows little divergence
__________ Dilates blood vessels to skeletal muscles

☞ Terminology Exercises

WORD PART	MEANING
af-	toward
astro-	star
corpor-	body
dendr-	tree
ef-	away from
encephal-	within the head, brain
esthes-	feeling
-fer-	to carry
gangli-	knot
gli-	glue
gloss-	tongue
lemm--	peel, rind
mening-	membrane
neur-	nerve
peri-	all around
pharyng-	throat
phas-	speech
pleg-	paralysis
plex-	interweave, network
sulc-	furrow, ditch

Use word parts given above or in previous chapters to form words that have the following definitions.

________________________ Pertaining to tongue and throat

________________________ Without feeling

________________________ Star-shaped cell

________________________ Inflammation of the brain

________________________ Without speech

Using the definitions of word parts given above or in previous chapters, define the following words.

Afferent ________________________

Neuroglia ________________________

Ganglion ________________________

Sulcus ________________________

Meningitis ________________________

Match each of the following definitions with the correct word.

________ Paralysis of four extremities

________ Network of nerves

________ Treelike processes

________ Four bodies

________ Membrane of a nerve fiber

A. Corpora quadrigemina

B. Dendrites

C. Neurilemma

D. Plexus

E. Quadriplegia

☞ Fun and Games

First, write the words that answer the clues in the blanks of the right-hand column, one letter per blank. Then, transfer each letter to the blank with the same number in the quotation by Juvenal at the bottom of the page.

Dura mater between cerebral hemispheres
29 66 43 0 74 14 37 70 26 54 5

Type of conduction on myelinated fibers
45 36 21 63 66 8 33 72 44

Type of cell that produces myelin
12 32 7 56 53 51 18

"Matter" that forms columns of spinal cord
20 69 40 39 14

Efferent process of a neuron
23 75 61 23 37 73 28 55

Posterior portion of the brain
74 30 54 25 26 47 3 59 27 34

"Matter" formed by myelinated fibers
31 7 5 68 42

Extent of sympathetic effects
4 50 23 70 15 65 72 22 1 23

Nonconducting tissue of nervous system
18 70 27 54 11 24 2 57 38

Junction between two neurons
6 67 51 1 46 13 75

Middle portion of the brainstem
35 9 61 16

Central region of the cerebellum
41 14 54 34 60 16

Collection of nerve cell bodies in PNS
52 48 18 62 43 73 64 51

Lobe that contains visual cortex
19 32 74 5 10 50 8 1 58

Innermost layer of meninges
71 73 66 34 48 28 70 37

QUOTATION:

1 2 3 4 5 6 7 8 9 10 11 12 13 14 15 16

K
17 18 19 20 21 22 23 24 25 26 27 28 29 30 31,

K
32 33 34 35 36 37 38 39 40 41 42 43 44 45 46 47 48 49 50 51 52,

53 54 55 56 57 58 59 60 61 62 63 64 65 66 67 68 69 70

71 72 73 74 75. Juvenal

9 Sensory System

☞ Chapter Outline/Objectives

Receptors and Sensations

1. Distinguish between general senses and special senses.
2. Classify sense receptors into five groups.
3. Explain what is meant by sensory adaptation.

General Senses

4. Describe the sense receptors for touch, pressure, proprioception, temperature, and pain.

Gustatory Sense

5. Describe the structure of a taste bud as it relates to the sense of taste.
6. Identify and locate the four different taste sensations and follow the impulse pathway from stimulus to the cerebral cortex.

Olfactory Sense

7. Locate the sense receptors for smell and trace the impulse pathway to the cerebral cortex.
8. Discuss the relationship between the sense of taste and sense of smell.

Visual Sense

9. List the bones that form the orbit of the eye.
10. Describe the protective features of the eye.
11. Describe the structure of the bulbus oculi and the significance of each component.
12. Explain the concept of light refraction and name the four refractive surfaces and media.
13. Explain how the eyes accommodate for near vision.
14. Identify the photoreceptor cells in the retina and describe the mechanism by which nerve impulses are triggered in response to light.
15. Trace the pathway of a visual impulse from where it is initiated to where it is interpreted in the cerebral cortex.

Auditory Sense

16. Distinguish between the outer ear, middle ear, and inner ear and describe the structure of each region.
17. Distinguish between the bony labyrinth and membranous labyrinth of the inner ear.
18. Describe the contribution each region of the ear makes to the sense of hearing.
19. Summarize the sequence of events in the initiation of auditory impulses.
20. Explain how differences in pitch and loudness are perceived.
21. Trace the pathway of auditory impulses from where they are initiated to where they are interpreted in the cerebral cortex.

Sense of Equilibrium

22. Distinguish between static equilibrium and dynamic equilibrium.
23. Identify and describe the structure of the components of the ear involved in static equilibrium and those involved in dynamic equilibrium.
24. Summarize the events in the initiation of impulses for static equilibrium and for dynamic equilibrium.
25. Identify the cranial nerve that transmits impulses for static and dynamic equilibrium to the cerebral cortex.

☞ Learning Exercises

Receptors and Sensations (Objectives 1-3)

1. Senses with receptors that are widely distributed within the body are called ________ senses. If the receptors are localized in a specific region, they are called _________.

2. In the blank at the left, write the type of sense receptor that responds to the indicated stimulus.

   ______________________ Light energy

   ______________________ Changes in pressure or movement

   ______________________ Temperature changes

   ______________________ Tissue damage

   ______________________ Changes in chemical composition

3. ____________________________ occurs when certain receptors are continually stimulated and no longer respond unless the stimulus becomes more intense.

General Senses (Objective 4)

1. Place a check (✓) in the space before each correct association.

   _______ Pacinian corpuscles - mechanoreceptors

   _______ Mechanoreceptors - proprioception

   _______ Meissner's corpuscles - heavy pressure

   _______ Pain - nociceptors

   ______ Golgi tendon organs - touch

2. Circle the word in the parentheses that makes the statement true.

   (Thermoreceptors/Nociceptors) exhibit rapid sensory adaptation.

   (Proprioceptors/Nociceptors) may send impulses after the stimulus is removed.

Gustatory Sense (Objectives 5 and 6)

1. Write the terms that best fit the following descriptive phrases.

   ____________________________ Another name for sense of taste

   ____________________________ Projections on tongue that contain taste buds

   ____________________________ Structures that project through a taste pore

   ____________________________ Type of receptors involved in taste

2. Write the terms that best fit the following descriptive phrases about taste.

   ____________________________ Four types of taste sensations

   ____________________________ Transmits impulses from anterior tongue

   ____________________________ Transmits impulses from posterior tongue

   ____________________________ Location of sensory cortex for taste in brain

Olfactory Sense (Objectives 7 and 8)

1. The chemoreceptors for olfaction are found in the olfactory epithelium located in the ______________________.
2. The olfactory nerve, cranial nerve number ________ transmits impulses to the olfactory cortex in the ____________________ lobe of the brain.
3. Airborne molecules may stimulate the sense of ___________ and the sense of ________________ at the same time.

Visual Sense (Objectives 9-15)

1. Write the terms that best fit the following descriptive phrases about the eye.

________________________ Gland that produces tears

________________________ Mucous membrane that lines the eyelid

________________________ White part of fibrous tunic

________________________ Anterior transparent part of fibrous tunic

________________________ Pigmented vascular tunic

________________________ Opening in the center of iris

________________________ Fluid in the anterior cavity of the eye

________________________ Muscles that regulate size of pupil

________________________ Muscles that control shape of the lens

________________________ Function of ciliary processes

________________________ Nervous tunic of the eye

________________________ Photoreceptor cells

________________________ Region where optic nerve penetrates eye

________________________ Depression in center of macula lutea

________________________ Gel-like fluid in posterior cavity

2. In the normal eye, light rays from distant objects are bent so they focus on the retina. This bending of light rays is called ___________________________.
3. Circle the correct word in each pair to make true statements.

   Accommodation is the adjustment needed to focus light rays for (distant/close) vision. To focus light rays from close objects, the ciliary muscle (relaxes/contracts), which (increases/decreases) the tension on the suspensory ligaments. When this happens, the lens becomes (thicker/thinner) and light rays are bent (more/less). In addition, the pupil (dilates/constricts).

4. Identify the listed structures of the bulbus oculi by matching them with the correct letters from the diagram. Color the two surfaces and two liquid media that bend light rays.

________ Anterior cavity

________ Choroid

________ Ciliary body

________ Cornea

________ Eyelid

________ Iris

________ Lens

________ Optic nerve

________ Posterior cavity

________ Retina

________ Sclera

________ Suspensory ligaments

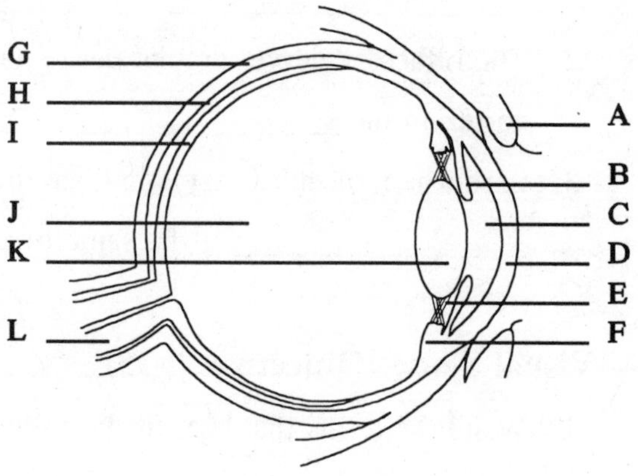

5. Write the terms that best fit the following descriptive phrases about photoreceptors.

________________________ Receptors for bright light vision

________________________ Pigment contained in rods

________________________ Receptors for color vision

________________________ Colors of light absorbed by color receptors

________________________ Receptors for dim light vision

________________________ Pigment that is very light sensitive

________________________ Receptors not present in the fovea

________________________ Vitamin required for photopigment synthesis

________________________ Area of retina lacking photoreceptors

________________________ Photoreceptors that exhibit greater convergence

6. Use the numbers 1 through 7 in the blanks to indicate the order of impulse transmission in the visual pathway. Use number 1 for where the impulse begins and number 7 for the final visual cortex.

________Optic chiasma ________Optic nerve ________Occipital lobe

________Optic tract ________Thalamus ________Photoreceptors

________Optic radiations

Auditory Sense (Objectives 16-21)

1. Identify the listed structures of the ear by matching them with the correct letters from the diagram. Use red to color the external ear, yellow for the middle ear, and blue for the inner ear.

____ Ampulla
____ Auditory tube
____ Auricle
____ Cochlea
____ Ext. auditory canal
____ Incus
____ Malleus
____ Semicircular canals
____ Stapes
____ Tympanic membrane
____ Vestibule
____ Vestibulocochlear nerve

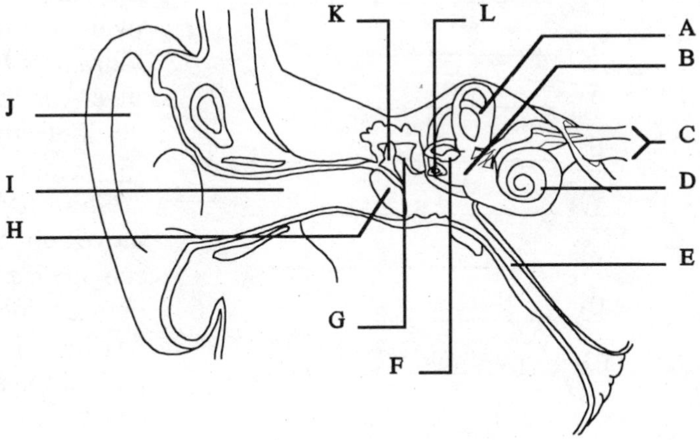

2. Write the terms that best fit the following descriptive phrases about the inner ear.

________________________ Series of interconnecting chambers in temporal bone
________________________ Membranous tubes inside bony labyrinth
________________________ Fluid inside the membranous labyrinth
________________________ Fluid outside the membranous labyrinth
________________________ Coiled portion of bony labyrinth
________________________ Coiled portion of membranous labyrinth
________________________ Membrane upon which the organ of Corti rests
________________________ Membrane between scala vestibuli and cochlear duct
________________________ Membrane between scala tympani and cochlear duct

3. Complete the following sentences about the events in the initiation of auditory impulses by writing the correct words in the spaces on the left.

1) ____________________
2) ____________________
3) ____________________
4) ____________________
5) ____________________
6) ____________________
7) ____________________
8) ____________________
9) ____________________
10) ____________________

Sound waves travel through the external auditory meatus until they reach the __1__, which starts to vibrate. These vibrations are transferred to the __2__, then the __3__, then the __4__, which is attached to the oval window. The oval window passes the vibrations to the __5__ in the scala vestibuli of the inner ear. As vibrations pass through the scala vestibuli, they create corresponding movement of the __6__ within the scala media. Finally, the vibrations are transferred to the basilar membrane and as it moves up and down, the __7__ of the organ of Corti rub against the __8__ and bend. This mechanical deformation initiates the nerve impulses that result in hearing. Pitch is determined by the region of the __9__ that vibrates in response to the sound. Loudness is determined by magnitude of __10__ of the membrane.

Sense of Equilibrium (Objectives 22-26)

1. Write the terms that best fit the following descriptive phrases about the sense of equilibrium.

____________________ Equilibrium of rotational or angular movement

____________________ Position of head relative to gravity

____________________ Receptor organ for static equilibrium

____________________ Two chambers containing static equilibrium organs

____

____________________ Calcium carbonate crystals on surface of the macula

____________________ Receptor organ for dynamic equilibrium

____________________ Location of dynamic equilibrium receptor organ

☞ Chapter Self-Quiz

1. Each of the following sensations is preceded by two blanks. In the first blank, indicate whether the sensation is one of the general senses or one of the special senses. In the second blank, indicate the type of receptor involved. Use the letter key that is given.

_____ _____ Vision
_____ _____ Touch
_____ _____ Taste
_____ _____ Smell
_____ _____ Cold shower
_____ _____ Toothache
_____ _____ Static equilibrium
_____ _____ Pressure
_____ _____ Pain from a burn
_____ _____ Hearing

G = General sense
S = Special sense

C = Chemoreceptor
M = Mechanoreceptor
N = Nociceptor
P = Photoreceptor
T = Thermoreceptor

2. Which of the following is <u>not</u> true about the sense of taste? (a) it is also called the gustatory sense; (b) in order to be tasted, substances must be dissolved; (c) receptors for the different tastes are randomly distributed over the surface of the tongue; (d) the facial and glossopharyngeal nerves transmit impulses for taste to the parietal lobe; (e) the receptors for taste are located in the taste buds

3. Which of the following is <u>not</u> true about protective features of the eye? (a) the frontal, ethmoid, and zygomatic bones contribute to the bony socket for the eye; (b) the gland that produces tears is located along the upper medial margin of the eye; (c) tears contain an enzyme that helps destroy pathogenic bacteria; (d) the eyes are opened by contraction of the levator palpebrae superioris muscle; (e) sebaceous glands are associated with the eyelashes

4. Which of the following represents the correct sequence for the visual pathway? (a) optic tract, optic chiasma, optic nerve, thalamus, optic radiations, visual cortex; (b) optic radiations, optic tract, optic chiasma, optic nerve, thalamus, visual cortex; (c) optic nerve, thalamus, optic tract, optic chiasma, optic radiations, visual cortex; (d) optic nerve, optic chiasma, optic tract, thalamus, optic radiations, visual cortex; (e) optic tract, thalamus, optic nerve, optic chiasma, optic radiations, visual cortex

5. Name the following:

____________________	Innermost tunic of the eyeball
____________________	White portion of the outer tunic
____________________	Fluid in the anterior cavity of the eye
____________________	Muscle that contracts to make the lens more convex
____________________	Center depression in the macula lutea
____________________	Receptors for color vision
____________________	Contains muscles that change the size of the pupil
____________________	Region where the optic nerve penetrates the eye
____________________	First structure that refracts light entering the eye
____________________	Pigment in the rods

6. Indicate whether each of the following parts of the ear is involved in hearing (H), equilibrium (E), or both (B).

_____ Malleus, incus, stapes

_____ Membranous labyrinth

_____ Semicircular canals

_____ Cochlea

_____ Utricle and saccule

_____ Tectorial membrane

_____ Endolymph

_____ Hair cells

_____ Organ of Corti

_____ Vestibulocochlear nerve

7. Name the following organs located in the ear:

____________________	Contains the receptor cells for hearing
____________________	Contains the receptor cells for static equilibrium
____________________	Contains the receptor cells for dynamic equilibrium

☞ Terminology Exercises

WORD PART	MEANING	WORD PART	MEANING
audi-	to hear	meat-	passage
coch--	snail	ocul-	eye
fove-	pit	olfact-	smell
gust-	taste	op-	eye
irid-	iris	ophthalm-	eye
kerat-	cornea	ot-	ear
lacr-	tears	presby-	old
lith-	stone	scler-	hard
lute-	yellow	tympan-	drum
macul-	spot	vitre-	glass

Use word parts given above or in previous chapters to form words that have the following definitions.

________________________________ Study of hearing

________________________________ Ear stone

________________________________ Glassy fluid

________________________________ Pertaining to taste

________________________________ Like a snail

Using the definitions of word parts given above or in previous chapters, define the following words.

Olfactory ________________________________

Macula lutea ________________________________

Intraocular ________________________________

Tympanectomy ________________________________

Lacrimal ________________________________

Match each of the following definitions with the correct word.

__________	Hearing of old age	A. Fovea centralis
__________	Inflammation of the cornea	B. Iridectomy
__________	Surgical repair of the ear	C. Keratitis
__________	Surgical excision of a portion of the iris	D. Otoplasty
__________	Central pit	E. Presbyopia

☞ Fun and Games

Braille is a system of writing for the blind that uses characters made up of raised dots. Each letter of the alphabet is represented by a specific pattern of dots as indicated in the alphabet that is given below. By learning this alphabet, the blind person can read with his or her fingers by feeling the patterns of raised dots.

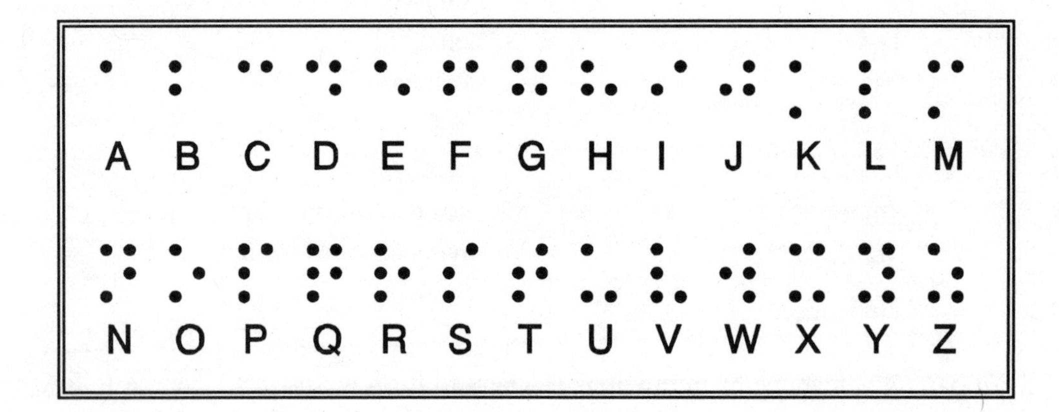

Use the braille alphabet given above to decipher this quotation by Ralph Waldo Emerson that reflects on the versatility of the eyes.

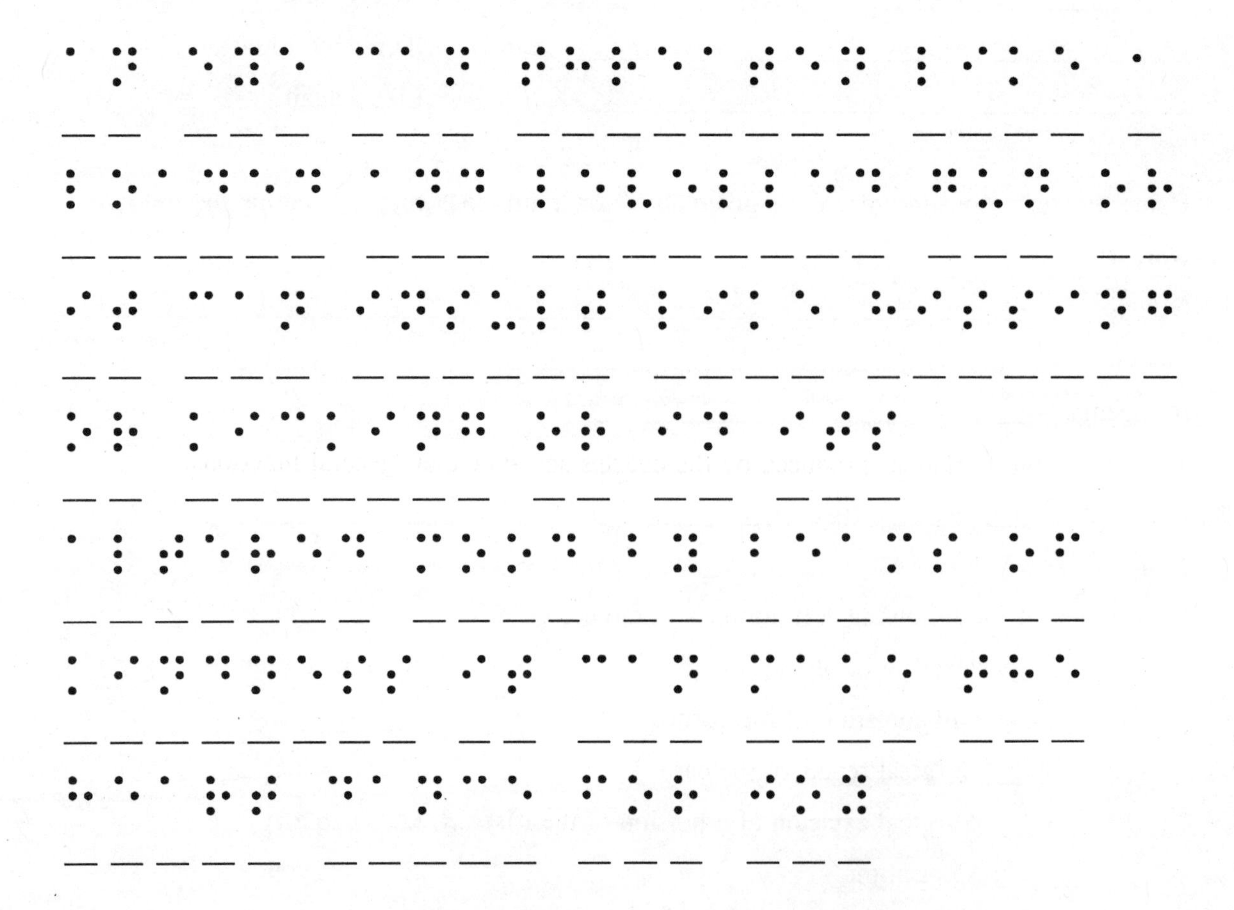

10 Endocrine System

☞ Chapter Outline/Objectives

Introduction to the Endocrine System

1. Compare the actions of the nervous system and the endocrine system.
2. Distinguish between the characteristics of exocrine glands and endocrine glands.

Characteristics of Hormones

3. Identify two chemical classes of hormones.
4. Explain why hormones affect only target tissues and not other tissues in the body.
5. Distinguish between receptor site locations for protein and steroid hormones.
6. Identify three different mechanisms for regulating hormone secretion.
7. Describe the mechanism of negative feedback regulation.

Endocrine Glands and Their Hormones

8. Identify the eight major endocrine glands.
9. Compare the anterior and posterior lobes of the pituitary gland in terms of embryonic origin, mechanisms that regulate their activity, and the hormones that are secreted.
10. Name six hormones secreted by the adenohypophysis and describe the action of each hormone.
11. Name two hormones secreted by the neurohypophysis and describe the action of each hormone.
12. Describe the location, structure, and hormones of the thyroid gland.
13. Discuss the location and actions of the parathyroid gland.
14. Compare the actions of calcitonin and parathyroid hormone.
15. Describe the location, structure, and regulation of the adrenal glands.
16. Identify the three groups of hormones that are secreted by the cortex of the adrenal gland and describe the functions of each group.
17. Discuss the physiologic effects of hypersecretion and hyposecretion of the adrenocortical hormones.
18. Name two hormones from the adrenal medulla and describe their effects.
19. Describe the location and structure of the pancreas.
20. Compare the sources and actions of glucagon and insulin.
21. Identify the principal androgen and state its general function.
22. Name two hormones produced by the ovaries and state their general functions.
23. Describe the location of the pineal gland and discuss its endocrine function.
24. Discuss the action of the thymus gland.
25. Name and describe the function of at least one hormone from the (a) gastric mucosa; (b) small intestine; (c) heart; and (d) placenta.

Prostaglandins

26. Differentiate between hormones and prostaglandins.

☞ Learning Exercises

Introduction to the Endocrine System (Objectives 1 and 2)

1. For each of the following phrases, place an N in the space provided if the the phrase pertains to the nervous system and an E if it pertains to the endocrine system.

_______ Acts through hormones

_______ Effect is localized

_______ Acts through electrical impulses

_______ Effect is of short-term duration

_______ Effect is generalized and long term

2. Exocrine glands have ____________ that carry the secretory product to a surface. In contrast, ____________________ are ductless and secrete their product directly into the ____________________ for transport to the target tissue.

Characteristics of Hormones (Objectives 3-7)

1. Write the terms that best fit the following descriptive phrases about the characteristics of hormones.

____________________________ Chemical nature of nonsteroid hormones

____________________________ Hormones that are lipid derivatives

____________________________ Cells that have hormone receptors

____________________________ Location of receptors for protein hormones

____________________________ Location of receptors for steroid hormones

____________________________ Three mechanisms for controlling secretion

____

____

2. Complete the following sentences about the reaction of a hormone with its receptor.

1) ____________________
2) ____________________
3) ____________________
4) ____________________
5) ____________________
6) ____________________
7) ____________________
8) ____________________
9) ____________________
10) ____________________

A __1__ hormone combines with a receptor on the surface of the cell. This activates an enzyme within the membrane, called __2__, and the enzyme diffuses into the cytoplasm where it catalyzes the production of __3__. This substance, a derivative of __4__, activates other enzymes that alter cellular activity. The __5__ is called the first messenger and __6__ is called the second messenger. A __7__ hormone diffuses through the cell membrane and reacts with a __8__ within the cytoplasm. The complex that forms enters the __9__ and reacts with __10__ to stimulate protein synthesis.

Endocrine Glands and Their Hormones (Objectives 8-25)

1. Identify the listed endocrine glands by matching them with the correct letters from the diagram.

_______ Adrenal gland

_______ Ovaries

_______ Pancreas

_______ Parathyroid

_______ Pineal gland

_______ Pituitary gland

_______ Thymus

_______ Thyroid

_______ Testes

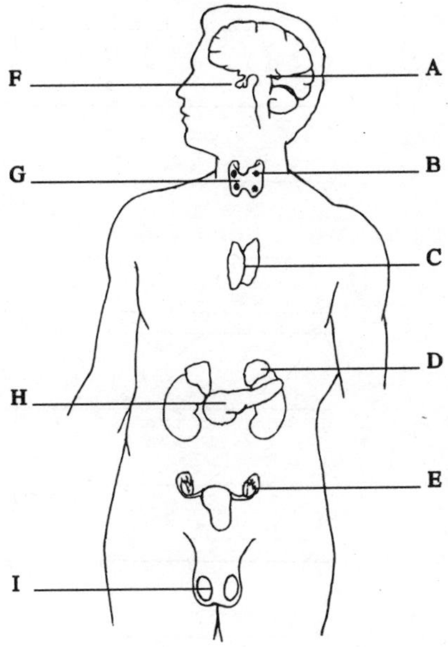

2. Match each of the following phrases with the region of the pituitary gland that it describes.

A = Adenohypophysis

B = Neurohypophysis

_______ Secretes ADH

_______ Stimulated by releasing hormones

_______ Secretes prolactin

_______ Derived from nervous tissue

_______ Secretes oxytocin

_______ Derived from embryonic oral cavity

_______ Secretes TSH

_______ Secretes growth hormone

_______ Secretes ACTH

_______ Regulated by nerve stimulation

3. Write the name of the pituitary hormones that best match the following functions.

_______________________ Stimulates secretion of thyroid hormone

_______________________ Stimulates the adrenal cortex

_______________________ Stimulates ovulation

_______________________ Stimulates uterine contractions

_______________________ Increases water reabsorption

_______________________ Stimulates spermatogenesis

_______________________ Stimulates protein synthesis and growth

_______________________ Stimulates ejection of milk

_______________________ Stimulates production of testosterone

_______________________ Stimulates milk production

4. Match each of the following phrases with the correct hormone from the thyroid and parathyroid glands.

C = Calcitonin
P = Parathyroid hormone
T = Thyroxine

_______ Secreted by parafollicular cells
_______ Requires iodine for production
_______ Increases blood calcium levels
_______ Secreted by thyroid follicles
_______ Increases rate of metabolism
_______ Reduces blood calcium levels
_______ Increases osteoclast activity
_______ Hyposecretion leads to nerve excitability

5. Write the terms that best fit the following descriptive phrases about the adrenal gland.

_______________ Region regulated by nerve impulses
_______________ Group of hormones from outer cortical layer
_______________ Primary mineralocorticoid hormone
_______________ Region regulated by negative feedback
_______________ Primary glucocorticoid
_______________ Increases blood sodium levels
_______________ Counteracts inflammatory response
_______________ Secreted by innermost layer of cortex
_______________ Two hormones from the adrenal medulla

6. Indicate whether each descriptive phrase refers to insulin or to glucagon by writing the appropriate word in the space provided.

_______________ Secreted in response to elevated blood glucose levels
_______________ Secreted by the alpha cells
_______________ Promotes breakdown of glycogen into glucose
_______________ Secreted by the beta cells
_______________ Stimulates liver to store glucose as glycogen
_______________ Secreted in response to low blood glucose levels
_______________ Increases blood glucose levels
_______________ Decreases blood glucose levels
_______________ Hyposecretion may result in diabetes mellitus

7. Each of the following hormones is preceded by two blanks. In the first blank, indicate the source of the hormone, and in the second blank, indicate a function of the hormone from the lists provided.

	Hormone	Source	Function
____ ____	Testosterone	A. Heart	1. Alkaline fluid from pancreas
____ ____	Melatonin	B. Ovaries	2. Contraction of gallbladder
____ ____	Thymosin	C. Pineal gland	3. Decrease blood volume
____ ____	Estrogen	D. Placenta	4. Development of lymphocytes
____ ____	HCG	E. Small intestine	5. Female sex characteristics
____ ____	Gastrin	F. Stomach	6. Maintains pregnancy
____ ____	Secretin	G. Testes	7. Male sex characteristics
____ ____	Progesterone	H. Thymus	8. Prepares uterus for pregnancy
____ ____	Cholecystokinin		9. Production of HCl in stomach
____ ____	Atrial peptin		10. Regulation of body rhythms

Prostaglandins (Objective 26)

1. Prostaglandins are similar to hormones, but they are different in many ways. They are derivatives of ________________________ and the cells that produce them are ________________________ throughout the body.

2. In contrast to hormones, the effects of prostaglandins are ________________________, ________________________, and ________________________.

☞ Chapter Self-Quiz

1. Classify the following hormones as either protein (P) or steroid (S) in chemical composition.

_____ Growth hormone

_____ Epinephrine

_____ Aldosterone

_____ Insulin

_____ Cortisol

_____ Follicle-stimulating hormone

2. Use the term "receptor sites" to write a sentence that defines target tissue.

3. Match each of the following hormones with the primary gland that secretes it.

A. Adenohypophysis
B. Adrenal cortex
C. Adrenal medulla
D. Neurohypophysis
E. Ovaries
F. Pancreas, alpha cells
G. Pancreas, beta cells
H. Parathyroid
I. Pineal
J. Testes
K. Thymus
L. Thyroid

____ Growth hormone
____ Gonadotropins
____ Epinephrine
____ Melatonin
____ Luteinizing hormone
____ Oxytocin
____ Aldosterone
____ Prolactin
____ Progesterone
____ Insulin
____ Calcitonin
____ Antidiuretic hormone
____ Cortisol
____ Glucagon
____ Testosterone
____ Thyroxine
____ TSH
____ Thymosin
____ Estrogens
____ ACTH

4. Name the hormone that is responsible for each of the following effects.

_________________ Promotes the development of sperm

_________________ Stimulates the production of milk

_________________ Plays a role in the development of immunity

_________________ Affects the body's rate of metabolism

_________________ Promotes the production of progesterone

_________________ Increases blood calcium levels

_________________ Regulates circadian rhythms

_________________ Causes uterine contractions

_________________ Promotes production of estrogens

_________________ Decreases blood glucose levels

_________________ Promotes kidney reabsorption of sodium ions

_________________ Responds in stress or emergency situations

_________________ Has an anti-inflammatory effect

_________________ Promotes the production of testosterone

_________________ Decreases blood calcium levels

5. Indicate whether each of the following pertains to prostaglandins (P) or hormones (H).

_____ Derived from arachidonic acid

_____ Transported to distant sites in the blood

_____ May be stored in the body

_____ Produced by cells widely distributed in the body

_____ Have a localized effect

_____ Derived from proteins and lipids

☞ Terminology Exercises

WORD PART	MEANING	WORD PART	MEANING
ad-	toward	lact-	milk
aden-	gland	oxy-	swift, rapid
-agon	assemble, gather together	para-	beside
andr-	male, maleness	pin-	pine cone
cortic-	outer region, cortex	-ren-	kidney
crin-	to secrete	ster-	steroid
di-	passing through	test-	eggshells, eggs
dips-	thirst	-toc-	birth
-gen-	to produce	trop-	to change, influence
-gest-	to carry, pregnancy	-uria	urine condition

Use word parts given above or in previous chapters to form words that have the following definitions.

_________________________ Substance for quick birth

_________________________ Like a pine cone

_________________________ Toward the kidney

_________________________ Steroid before pregnancy

_________________________ Passing through urine

Using the definitions of word parts given above or in previous chapters, define the following words.

Adenoma _________________________

Endocrinology _________________________

Testosterone _________________________

Parathyroid _________________________

Adrenocorticotropin _________________________

Match each of the following definitions with the correct word.

__________ Assembles glucose into the blood

__________ Surgical excision of the adrenal gland

__________ Excessive thirst

__________ Hormones that produce males or maleness

__________ Hormone necessary before milk production

A. Adrenalectomy

B. Androgens

C. Glucagon

D. polydipsia

E. prolactin

☞ Fun and Games

This is a variation of the word game "Hangman." Guess any letter for the first word and find the number that corresponds to that letter in the Letter Chart at the bottom of the page. Then find that same number above the line that divides each cell in the Position Chart on the right. If the letter you guessed appears in the word, its position is given by the number or numbers below the line that divides each cell in the Position Chart. If the letter does not appear in the word, 0 will be indicated under the line. If the letter you guessed does not appear in the word, start drawing a "stick person" on a gallows--first a head, then a body, followed by two arms and two legs. You are allotted six wrong guesses before you are "hanged." Clue: All words are hormones.

WORDS		LETTERS MISSED
#1	1 2 3 4 5 6 7 8 9 10 11 12	|_|_|_|_|_|_|
#2	1 2 3 4 5 6 7 8 9	|_|_|_|_|_|_|
#3	1 2 3 4 5 6 7 8	|_|_|_|_|_|_|
#4	1 2 3 4 5 6 7 8 9 10	|_|_|_|_|_|_|
#5	1 2 3 4 5 6 7 8 9 10 11	|_|_|_|_|_|_|
#6	1 2 3 4 5 6 7 8 9 10 11 12	|_|_|_|_|_|_|
#7	1 2 3 4 5 6 7	|_|_|_|_|_|_|
#8	1 2 3 4 5 6 7 8	|_|_|_|_|_|_|
#9	1 2 3 4 5 6 7 8	|_|_|_|_|_|_|
#10	1 2 3 4 5 6 7 8 9 10 11 12	|_|_|_|_|_|_|

POSITION CHART

1 / 6	2 / 9	3 / 9	4 / 7	5 / 1, 6
6 / 2	7 / 1, 5	8 / 1	9 / 8	10 / 2,8,12
11 / 4	12 / 1, 7	13 / 1, 4	14 / 3, 6	15 / 7, 10
16 / 0	17 / 1	18 / 7	19 / 4	20 / 7
21 / 8, 3	22 / 0	23 / 2	24 / 11	25 / 4, 9
26 / 0	27 / 8	28 / 2	29 / 2, 9	30 / 9
31 / 0	32 / 8	33 / 5	34 / 4	35 / 2
36 / 1, 6	37 / 4	38 / 0	39 / 5, 9	40 / 0
41 / 5	42 / 6	43 / 4	44 / 6	45 / 6
46 / 4	47 / 0	48 / 11	49 / 0	50 / 3
51 / 12	52 / 3	53 / 5	54 / 8, 10	55 / 10
56 / 7	57 / 0	58 / 1, 4, 7	59 / 3	60 / 3, 10
61 / 6	62 / 2, 7	63 / 3	64 / 8	65 / 3
66 / 8	67 / 0	68 / 0	69 / 7, 11	70 / 6
71 / 1	72 / 1	73 / 7	74 / 5	75 / 2, 4
76 / 5, 8,12	77 / 5	78 / 5	79 / 6	80 / 3
81 / 11	82 / 5, 10	83 / 2	84 / 3	85 / 2

LETTER CHART

	A	B	C	D	E	F	G	H	I	J	K	L	M	N	O	P	Q	R	S	T	U	V	W	X	Y	Z
#1	75	16	68	38	51	26	67	42	22	47	68	31	3	81	15	72	40	21	68	33	49	16	38	57	26	47
#2	41	38	79	22	31	40	49	67	66	16	31	11	47	2	59	17	26	85	16	20	38	67	26	49	16	31
#3	16	26	70	38	47	67	22	38	4	49	68	26	40	27	7	67	57	47	38	46	22	40	68	35	84	26
#4	83	57	13	68	16	57	16	31	39	47	67	65	22	54	73	26	38	31	47	1	16	26	31	47	57	22
#5	71	22	31	50	69	38	49	68	26	40	57	6	16	55	25	49	68	32	77	45	31	49	22	67	31	68
#6	40	49	67	16	10	31	47	67	22	38	49	68	26	24	82	47	57	30	14	58	26	38	16	57	22	38
#7	26	38	47	57	68	40	57	16	5	31	47	74	67	62	16	22	38	40	52	22	19	22	31	40	49	67
#8	78	16	34	31	47	22	36	38	49	68	26	23	40	64	56	26	31	57	16	49	63	26	38	47	68	16
#9	67	22	38	49	12	57	61	16	31	47	67	22	38	9	53	38	67	43	28	80	16	22	31	40	49	26
#10	68	26	40	57	76	49	37	68	26	40	57	16	38	48	60	8	16	29	44	18	38	47	57	26	67	38

11 Blood

☞ Chapter Outline/Objectives

Functions and Characteristics of Blood

1. Describe five physical characteristics of blood.
2. List six functions of blood.

Composition of Blood

3. Identify the two parts of a blood sample and state the normal percentage of total volume for each one.
4. Describe the composition of blood plasma.
5. List three categories of formed elements in the blood.
6. Identify seven formed elements of the blood and state at least one function for each formed element.
7. Discuss the life cycle of erythrocytes.
8. Differentiate between five types of leukocytes and tell whether each one is an agranulocyte or a granulocyte.

Hemostasis

9. Describe the three processes that constitute hemostasis.
10. Summarize the series of chemical reactions involved in the formation of a blood clot into three main steps.
11. Define fibrinolysis.

Blood Typing and Transfusions

12. Define agglutinogen and agglutinin.
13. State the agglutinogens and agglutinins present in each of the four ABO blood types and explain why different blood types are incompatible for transfusions.
14. Explain the difference between Rh+ and Rh− blood.
15. Discuss the pathogenesis and treatment of hemolytic disease of the newborn.

☞ Learning Exercises

Functions and Characteristics of Blood (Objectives 1 and 2)

1. Write the terms that best fit the following phrases about the functions and characteristics of blood.

______________________	pH range of the blood
______________________	Normal volume of blood
______________________	Types of substances transported by blood (3)
______________________	
______________________	
______________________	Three regulatory functions of blood
______________________	
______________________	

Composition of Blood (Objectives 3-8)

1. The following statement is false. Rewrite the statement to make it true.

   Blood is 70% plasma and 30% formed elements.

2. Complete the following table about plasma proteins.

Plasma Protein	Percent of Total	Function
	60%	
Globulin		
		Blood clotting

3. Give two examples for each of the following plasma components.

   Nonprotein molecules that contain nitrogen

   Nutrients

   Gases

   Electrolytes

4. Write the scientific term for each of the three types of formed elements in the blood.

   Red blood cells:

   White blood cells:

   Platelets:

5. Identify each of the formed elements illustrated in the following diagram by matching the letters with the names. Color the granulocytes yellow and color the agranulocytes green.

   _______ Basophil

   _______ Eosinophil

   _______ Erythrocyte

   _______ Lymphocyte

   _______ Monocyte

   _______ Neutrophil

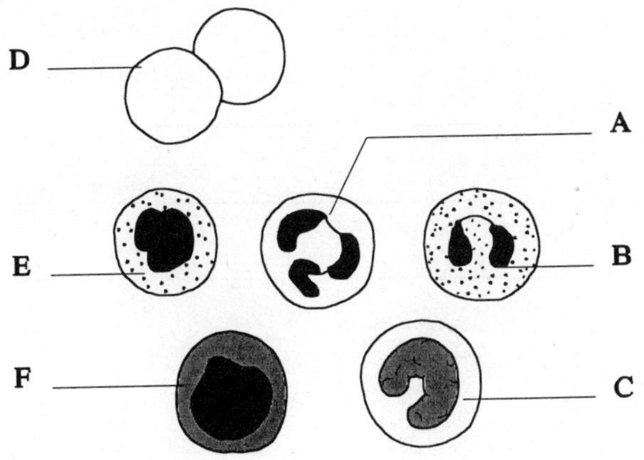

D

A

E

B

F

C

6. Write the terms that best fit the following phrases about the formed elements of blood.

_______________ Anucleate biconcave discs

_______________ Immature RBCs circulating in the blood

_______________ Normal range of RBCs/mm^3 in blood

_______________ Large protein pigment in RBCs

_______________ Process by which WBCs enter tissue spaces

_______________ Normal range of WBCs/mm^3 in blood

_______________ Most numerous leukocyte

_______________ Leukocyte with red-staining granules

_______________ Contains histamine and heparin

_______________ Largest leukocyte

_______________ Function in antibody production

_______________ Multilobed nucleus

_______________ Called macrophages in the tissues

_______________ Phagocytic granulocyte

_______________ Phagocytic agranulocyte

_______________ Called mast cells in the tissues

_______________ Help counteract effects of histamine

_______________ Function of thrombocytes

_______________ Normal range of platelets/mm^3 in the blood

_______________ Cell that breaks apart to form platelets

7. Write the terms that best fit the following phrases about erythrocyte production.

____________________	Stem cell in the bone marrow
____________________	Hormone that stimulates RBC production
____________________	Mineral necessary for healthy RBCs
____________________	Two vitamins necessary for RBC production
____________________	
____________________	Necessary for absorption of vitamin B_{12}
____________________	Average life span of RBCs
____________________	Organs where old RBCs are phagocytized
____________________	
____________________	Waste product of RBC destruction

Hemostasis (Objectives 9-11)

1. Write the terms that best fit the following phrases about hemostasis. production.

____________________	Initial reaction in hemostasis
____________________	Process of collagen attracting platelets
____________________	Vitamin necessary for clot formation
____________________	Mineral necessary for clot formation
____________________	Dissolution of a clot

2. The following diagram shows three reactions in blood clotting. Identify substances A, B, C, and D.

Damaged Tissues → Formation of Substance A

Substance B → Substance C

Substance D → Fibrin

Substance A____________________ Substance C____________________

Substance B____________________ Substance D____________________

Blood Typing and Transfusions (Objectives 12-15)

1. Complete the following table about the different ABO blood types. For the permissible donor column, assume that a person with the given blood type needs a transfusion and the preferred type is not available. What are the other possible donor types? For the permissible recipient column, what blood types may receive the indicated type if the preferred type is not available?

Blood Type	Agglutinogen	Agglutinin	Permissible Donors	Permissible Recipients
A				
	B			
		None		
	None			

2. Write the terms that best fit the following phrases about Rh+ and Rh− blood.

________________________ Agglutinogen in Rh+ blood

________________________ Agglutinogen in Rh− blood

________________________ Agglutinin in unsensitized Rh+ blood

________________________ Agglutinin in unsensitized Rh− blood

3. Answer the following questions about Rh+ and Rh− blood.

What happens if an unsensitized Rh− individual receives Rh+ blood?

What happens if a sensitized Rh− individual receives Rh+ blood?

An Rh+ individual does not develop anti-Rh antibodies after receiving Rh− blood. Why?

What maternal and fetal blood types may lead to hemolytic disease of the newborn?

Maternal type: Fetal type:

☞ Chapter Self-Quiz

1. Which one of the following is not true about blood? (a) the normal volume of blood is about 5 liters; (b) the normal pH is slightly acidic; (c) blood helps regulate body temperature; (d) blood is classified as a connective tissue; (e) blood is more viscous than water

2. Which of the following statements about plasma is not true? (a) plasma is about 90% water; (b) fibrinogen is a plasma protein that functions in clotting; (c) the most abundant plasma proteins are the globulins; (d) bicarbonate ions are found as solutes in the plasma; (e) some oxygen and carbon dioxide are transported as solutes in the plasma

3. Write the terms that best fit each of the following.

_______________________ Production of formed elements

_______________________ The stem cell from which blood cells develop

_______________________ Hemoglobin that is combined with oxygen

_______________________ Most numerous type of leukocyte

_______________________ Hormone that stimulates RBC production

_______________________ Process of WBCs moving through capillary walls

_______________________ Leukocyte that produces antibodies

4. Indicate whether each of the following pertains to erythrocytes (E), leukocytes (L), or thrombocytes (T).

_____ Fragments of large cells
_____ Contain hemoglobin
_____ May have granules in the cytoplasm
_____ Some are phagocytic
_____ Biconcave discs
_____ Platelets
_____ Transport oxygen
_____ Primary function is blood clotting
_____ Function in prevention of disease
_____ Normal number is about 5 million/mm^3

5. Number the following events of hemostasis in the sequence in which they occur. The first event is 1 and the final event is 5.

_____ Prothrombin is converted to thrombin

_____ Smooth muscle in vessel walls contracts

_____ Formation of prothrombin activator

_____ Fibrinogen is converted to fibrin

_____ Collagen attracts platelets to form platelet plug

6. Indicate whether each of the following pertains to blood type A, B, AB, or O. Some may have more than one answer.

__________ Has anti-A agglutinins

__________ Can be given to type B individuals

__________ Has no agglutinogens

__________ Universal recipient

__________ Has no agglutinins

__________ Can be given to type O individuals

__________ Can donate to type A individuals

__________ Has type B agglutinogens

__________ Universal donor

__________ Reacts with type AB donor blood

7. Which of the following statements about Rh blood types is not true? (a) about 85% of the population is Rh+; (b) normally, Rh− individuals do not have anti-Rh agglutinins; (c) normally, Rh+ individuals do not have anti-Rh agglutinins; (d) normally, Rh− individuals have Rh agglutinogens; (e) normally, Rh+ individuals have Rh agglutinogens

8. Indicate whether each of the following statements is **true** or **false**.

_____ When Rh− blood is given to an Rh+ individual, the Rh+ individual develops anti-Rh agglutinins.

_____ When Rh+ blood is given to an Rh− individual, the Rh- individual develops anti-Rh agglutinogens.

_____ Hemolytic disease of the newborn may develop when the mother is Rh− and the fetus is Rh+

_____ Hemolytic disease of the newborn causes agglutination and hemolysis of the fetal blood.

_____ If hemolytic disease of the newborn develops, the fetus should receive a transfusion of Rh+ blood.

☞ Terminology Exercises

WORD PART	MEANING	WORD PART	MEANING
agglutin-	clumping, sticking together	leuk-	white
anti-	against	-lysis	destruction
coagul-	clotting	mon-	single, one
-emia	blood condition	-penia	deficiency, lack of
erythr-	red	-phil	love, affinity for
fibr-	fiber	-poiesis	formation of
glob-	globe	-rrhage	burst forth, flow
hem-	blood	-stasis	control
kary-	nucleus	thromb-	clot

Use word parts given above or in previous chapters to form words that have the following definitions.

_______________________ Destruction of blood

_______________________ Deficiency of clotting cells

_______________________ Formation of blood

_______________________ Affinity for basic dye

_______________________ Destruction of fibrin

Using the definitions of word parts given above or in previous chapters, define the following words.

Agglutination _______________________

Anticoagulant _______________________

Fibrinogen _______________________

Leukocyte _______________________

Thrombocyte _______________________

Match each of the following definitions with the correct word.

_______ Substance before thrombin

_______ Globe-shaped protein in blood

_______ Condition of too many cells in blood

_______ Flowing of blood

_______ Cell with large nucleus

A. hemoglobin
B. hemorrhage
C. megakaryocyte
D. polycythemia
E. prothrombin

☞ Fun and Games

Each of the following terms is taken from this chapter and pertains to the blood in some way. The words also contain people's names with the letters in the correct sequence but there may be additional letters between the letters of the name. Fill in the spaces around the thirty names to complete the terms.

P _ A _ M _

F O R _ _ D   _ _ _ _ _ _ _

_ R _ _ _ _ O _ Y _ _

_ L O _ U _ _ _

E _ _ _ _ R _ _ _ I _ _ _ N

_ _ _ _ _ T I N A _ _ _ _

L _ U _ _ C Y _ _

_ _ S T A _ _ N _

_ _ B _ _ _ _ _ E N

_ _ _ _ _ _ _ _ _ M I A

_ G _ L _ _ _ _ _ _ E N

_ _ R O _ B _ _

L _ _ _ _ O _ _ T _

D _ A _ _ D E _ _ _

M A _ R _ _ _ _ G E

_ _ _ _ _ _ P H I L

A _ _ _ _ _ _ N _ N

_ R O _ _ _ _ _ B I N

N E _ _ _ _ _ _ I L

M A _ _   C _ _ _

P _ A T _ _ _ _

_ _ S O P H I _

B _ _ _ R U _ I N

M _ _ A _ _ R _ _ C Y _ _

G R A   _ _ _ _ C _ _ E

_ _ M A _ _ _ R I _

T _ _ O M _ _ _ _ _ _

A L _ _ _ _ _

A _ R _ _ _ _ _ _ _ T _

_ A _ _ _   _ L _ B _ _ I N

12 Heart

☞ Chapter Outline/Objectives

Overview of the Heart

1. Describe the size and location of the heart.
2. Describe the pericardium and pericardial cavity.

Structure of the Heart

3. Identify the layers of the heart wall and state the type of tissue in each layer.
4. Discuss the structure and function of each of the four chambers of the heart.
5. Identify the valves associated with the heart and describe their location, structure, and function.
6. Label a diagram of the heart, identifying the chambers, valves, and associated vessels.
7. Trace the pathway of blood flow through the heart, including chambers, valves, and pulmonary circulation.
8. Identify the major vessels that supply blood to the myocardium and return the deoxygenated blood to the right atrium.

Physiology of the Heart

9. Describe the components and function of the conduction system of the heart.
10. Correlate the deflections on an ECG with the electrical events in the conduction system.
11. Define systole and diastole.
12. Summarize the events of a complete cardiac cycle.
13. Correlate the heart sounds heard with a stethoscope with the events of the cardiac cycle.
14. Explain what is meant by cardiac output and describe the factors that affect it.
15. Describe the pathway be which the central nervous system regulates heart rate.

☞ Learning Exercises

Overview of the Heart (Objectives 1 and 2)

1. All of the following statements are false. Rewrite each statement to make it true.

   The base of the heart is directed inferiorly and to the left.

   About 1/3 of the heart mass is on the left side.

   The heart is located in the anterior mediastinum, between the second and fifth ribs.

2. Arrange the following coverings around the heart in the correct sequence by numbering them in the order in which they occur. Start with 1 for the outermost layer and proceed to 4 for the innermost layer.

   _______ Epicardium　　　_______ Fibrous pericardium

   _______ Pericardial cavity　　　_______ Parietal pericardium

Structure of the Heart (Objectives 3-8)

1. Write the terms that best fit the following phrases about the structure of the heart.

______________________	Middle layer of the heart wall
______________________	Chamber that receives blood from SVC
______________________	Valve between the right atrium and ventricle
______________________	Inner lining of the heart wall
______________________	Receives oxygenated blood from the lungs
______________________	Thin region of the interatrial septum
______________________	Receives blood from the right atrium
______________________	Ridges of myocardium in ventricles
______________________	Atrioventricular valve on left side of heart
______________________	Valve at base of the aorta
______________________	Arteries that supply blood to the myocardium
______________________	Chamber that receives blood from IVC

2. Identify the listed structures of the heart by matching them with the correct letters from the diagram. Use blue to color the chambers and vessels that contain oxygen-poor blood. Use red to color the chambers and vessels that contain oxygen-rich blood.

______ Right atrium
______ Right ventricle
______ Left atrium
______ Left ventricle
______ Interventricular septum
______ Superior vena cava
______ Inferior vena cava
______ Pulmonary trunk
______ Ascending aorta
______ Right brachiocephalic vein
______ Brachiocephalic artery
______ Aortic semilunar valve
______ Pulmonary semilunar valve
______ Tricuspid valve
______ Bicuspid valve
______ Left common carotid artery
______ Left subclavian artery
______ Left pulmonary artery

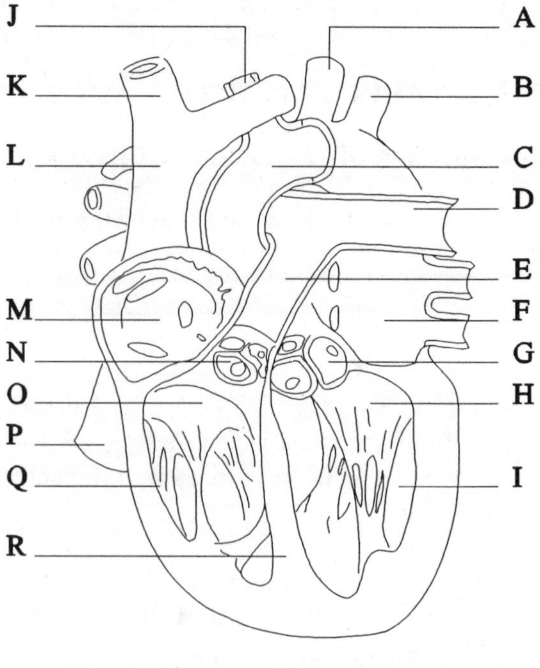

J K L M N O P Q R A B C D E F G H I

3. Trace the pathway of blood through the heart by numbering the following structures in the correct sequence. Start with number 1 for the venae cavae.

____ Aortic semilunar valve
____ Ascending aorta
____ Bicuspid valve
____ Capillaries of lungs
____ Left atrium
____ Left ventricle
____ Pulmonary semilunar valve
____ Pulmonary trunk
____ Pulmonary veins
____ Pulmonary arteries
____ Right atrium
____ Right ventricle
____ Tricuspid valve
____ Venae cavae

Physiology of the Heart (Objectives 9-15)

1. Write the terms that best fit the following phrases about the conduction system of the heart and the cardiac cycle.

____________________ Pacemaker of the heart
____________________ Length of cardiac cycle at 72 beats/minute
____________________ Indicates depolarization of atria on ECG
____________________ Transmits impulses to atria and AV node
____________________ A recording of the electrical activity of heart
____________________ Indicates depolarization of ventricles on ECG
____________________ Contraction phase of the ventricles
____________________ Carry impulses to myocardium
____________________ Indicates repolarization of ventricles on ECG
____________________ Relaxation phase of the ventricles
____________________ Transmit impulses away from AV node
____________________ Length of atrial systole in one cardiac cycle
____________________ Normally has fastest rate of depolarization
____________________ Length of ventricular systole in one cycle

2. In the space before each phrase about events in the cardiac cycle, write S if the phrase refers to ventricular systole and write D if it refers to ventricular diastole.

____ Atria are in systole

____ Atrioventricular (AV) valves open

____ Atrioventricular (AV) valves close

____ Semilunar (SL) valves open

____ Semilunar (SL) valves close

____ Blood is ejected from the heart

____ Pressure in ventricles decreases

____ Blood enters the ventricles

3. Heart sounds are associated with heart valves closing. Which valves are associated with the following sounds:

   a. First heart sound (lubb)

   b. Second heart sound (dupp)

4. In the space before each statement about factors that influence cardiac output, write T if the statement is true and write F if the statement is false. If the statement is false, change the underlined word(s) to make the statement true.

____ Cardiac output equals heart rate <u>plus</u> stroke volume.

____ Stroke volume is the amount of blood ejected from the heart during each <u>minute</u>.

____ Increased venous return <u>decreases</u> contraction strength.

____ Sympathetic stimulation <u>decreases</u> contraction strength.

____ Increased contraction strength <u>increases</u> cardiac output.

____ Parasympathetic nerves <u>decrease</u> heart rate.

____ Increased end-diastolic volume <u>decreases</u> stroke volume.

____ Sympathetic stimulation <u>increase</u> heart rate.

____ Increased end diastolic volume <u>increases</u> cardiac output.

____ Most changes in heart rate are mediated through the cardiac center in the <u>pons</u>.

____ Epinephrine <u>decreases</u> heart rate.

____ Increased carbon dioxide concentration in the tissues <u>increases</u> heart rate.

☞ Chapter Self-Quiz

1. Which one of the following is <u>not</u> true about the location of the heart? (a) it is located in the middle mediastinum; (b) 2/3 of the mass of the heart is on the left side; (c) the heart extends between the second and sixth ribs; (d) the base of the heart is the lowest portion; (e) the heart is about the size of a closed fist

2. **True or False**: The fibrous pericardium is the outermost layer of the heart wall.

3. Use the words epicardium, endocardium, and myocardium to describe the structure of the heart wall.

4. Indicate whether each of the following pertains to the right atrium (RA), left atrium (LA), right ventricle (RV), or left ventricle (LV). More than one response may apply.

   _____ Receives blood from pulmonary veins

   _____ Pumps blood to the lungs

   _____ Contains unoxygenated blood

   _____ Contains papillary muscles

   _____ Receives blood from the inferior vena cava

   _____ Pumps blood through the aortic semilunar valve

   _____ Pumps blood into systemic circulation

   _____ Receives blood through the tricuspid valve

   _____ Coronary sinus opens into this chamber

   _____ Bicuspid valve is at the exit of this chamber

5. Name the two major branches of the left coronary artery.

6. Indicate whether each of the following statements is true (T) or false (F).

_____ Atrial systole lasts longer than ventricular systole

_____ All chambers are in simultaneous diastole for half the cardiac cycle

_____ An increase in right ventricular pressure closes the tricuspid valve and opens the semilunar valve

_____ Most of the blood enters the ventricles during atrial systole

_____ The P wave on an electrocardiogram corresponds to atrial depolarization

_____ The first heart sound is heard when the semilunar valves open

_____ The T wave on an electrocardiogram corresponds to atrial repolarization

_____ Impulses are carried throughout ventricular myocardium by conduction myofibers

_____ Ventricular diastole lasts longer than atrial diastole

_____ Blood is oxygenated before it enters the right ventricle

7. If stroke volume is 65 ml and heart rate is 75 beats/minute, what is the cardiac output per minute?

8. Which of the following statements is <u>not</u> true about cardiac output? (a) increased venous return increases contraction strength; (b) increased parasympathetic stimulation increases cardiac output; (c) increased end-diastolic volume increases contraction strength; (d) increased end-diastolic volume increases cardiac output; (e) increased sympathetic stimulation increases heart rate

☞ Terminology Exercises

WORD PART	MEANING
aort-	lift up
atri-	entrance room
brady-	slow
cardi-	heart
cusp-	point
diastol-	expand, separate
lun-	moon shaped

WORD PART	MEANING
meg-	large
sphygm-	pulse
sten-	narrowing
systol-	contraction
tachy-	fast, rapid
valvu-	valve

Use word parts given above or in previous chapters to form words that have the following definitions.

______________________ Heart entrance chamber

______________________ Rapid heart beat

______________________ Has three points

______________________ Vessel that lifts up from heart

______________________ Half-moon shaped

Using the definitions of word parts given above or in previous chapters, define the following words.

Cardiomegaly ______________________

Mitral stenosis ______________________

Valvulitis ______________________

Bradycardia ______________________

Systole ______________________

Match each of the following definitions with the correct word.

________	Membrane around the heart	A. cardiomyopathy
________	Disease condition of heart muscle	B. electrocardiogram
________	Surgical repair of a valve	C. endocarditis
________	Inflammation of heart lining	D. pericardium
________	Recording of electrical activity of heart	E. valvuloplasty

☞ Fun and Games

Thirty-three terms relating to the heart are described below. Determine the term that fits each description, then place it in the correct position on the grid. The descriptions are arranged according to the number of letters in the correct response.

3 Letters
Heart activity record
Records ventricular contraction

4 Letters
Pointed end of the heart
End directed to the right
First heart sound
Second heart sound

5 Letters
Vessel exiting left ventricle
Return blood to the heart

6 Letters
Thin-walled chamber
Heart pacemaker
Left AV valve
Takes blood away from heart

7 Letters
Contraction phase of cycle
Atrial appendage

8 Letters
Narrowing of an opening
Relaxation phase of cycle
Left AV valve

9 Letters
Pumping chamber of heart
Right AV valve
Valve at exit from heart

10 Letters
Outer layer of heart wall
Cardiac muscle layer
Inflammation of a valve
L coronary artery branch

11 Letters
Slow heart rate
Surrounds the heart
Innermost layer of wall
Thin area of septum
Rapid heart rate

12 Letters
Listening to the heart
Enlarged heart
Ejected during systole
Inflammation of lining

13 Blood Vessels

☞ Chapter Outline/Objectives

Classification and Structure of Blood Vessels

1. Describe the structure and function of arteries, capillaries, and veins.

Physiology of Circulation

2. Discuss how oxygen, carbon dioxide, glucose, and water move across capillary walls.
3. Discuss three factors that affect blood flow in arteries.
4. State three actions that provide pressure gradients for blood flow in veins.
5. Identify at least six commonly used pulse points.
6. Distinguish between systolic pressure, diastolic pressure, and pulse pressure. State normal values for each.
7. Discuss four primary factors that affect blood pressure.
8. Explain the role of baroreceptors and chemoreceptors in the regulation of blood pressure.
9. Describe the renin/angiotensin/aldosterone mechanism for blood pressure regulation.

Circulatory Pathways

10. Trace blood through the pulmonary circuit from the right atrium to the left atrium.
11. Identify the major systemic arteries.
12. Describe the blood supply to the brain.
13. Identify the major systemic veins.
14. Describe five features of fetal circulation that make it different from adult circulation.

☞ Learning Exercises

Classification and Structure of Blood Vessels (Objective 1)

1. In the space before each phrase about blood vessels, write A if the phrase refers to an artery, C if it refers to a capillary, and V if it refers to a vein.

____ Carries blood away from the heart

____ Functions in the exchange of substances between blood and tissue cells

____ Wall consists of simple squamous epithelium

____ Carries blood toward the heart

____ Has valves

____ Has three relatively thick layers in the wall

____ Has three relatively thin layers in the wall

____ May have vasa vasorum

2. Name the tunic or layer in the wall of a blood vessel that contains predominantly

a. Connective tissue   b. Simple squamous epithelium   c. Smooth muscle

Physiology of Circulation (Objectives 2-9)

1. Identify the passive transport process by which each of the following substances moves through the capillary wall.

_______________ Carbon dioxide

_______________ Oxygen

_______________ Water (two processes)

____

2. Write an A before the phrases that pertain to the arterial end of a capillary and write a V before the phrases that pertain to the venule end of a capillary.

_______ Osmotic pressure is greater than hydrostatic pressure

_______ Hydrostatic pressure is greater than osmotic pressure

_______ Net movement of fluid out of the capillary

_______ Net movement of fluid into the capillary

3. In the spaces on the left, write the answers that match the statements about blood flow and blood pressure.

_______________ Vessel with the greatest pressure

_______________ Type of vessel with greatest area

_______________ Type of vessel with the lowest pressure

_______________ Type of vessel with the slowest blood flow

_______________ Vessel with the fastest blood flow

_______________ Primary force that moves blood

_______________ Three actions that move blood through veins

____

____

_______________ Pressure during ventricular contraction

_______________ Pressure during ventricular relaxation

_______________ Instrument used to measure blood pressure

_______________ Systolic pressure minus diastolic pressure

_______________ Blood flow sounds heard in a stethoscope

_______________ Effect of ↑CO_2 on precapillary sphincters

4. Identify the pulse points indicated on the following diagram.

A. ______________________________

B. ______________________________

C. ______________________________

D. ______________________________

E. ______________________________

F. ______________________________

G. ______________________________

H. ______________________________

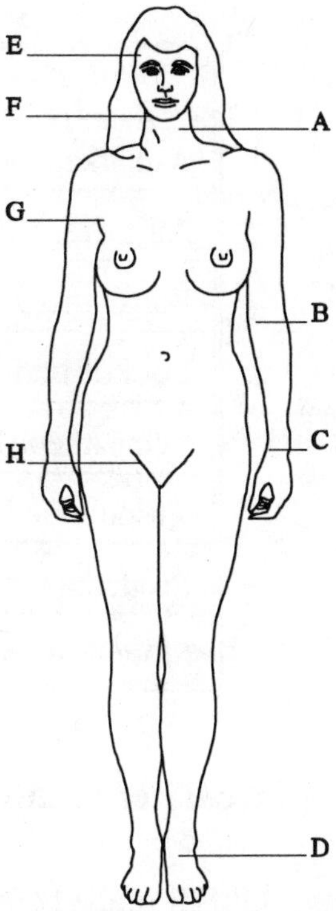

5. In the spaces on the left, write the answers that match the statements about the regulation of blood pressure.

______________________________ Location of vasomotor and cardiac centers

______________________________ Center that regulates heart rate

______________________________ Center that regulates blood vessel diameter

______________________________ Type of receptors that respond when vessel walls stretch

______________________________ Type of receptors that respond to carbon dioxide and hydrogen ion concentrations

______________________________ Substance produced by the kidneys that has a role in blood pressure regulation

______________________________ Two actions of angiotensin that increase blood pressure

____

______________________________ Hormone that conserves sodium ions to increase blood pressure

6. Place a check (✓) in the appropriate column to indicate whether the given condition tends to increase or decrease blood pressure.

Given Condition	Increase	Decrease
A marked increase in blood volume		
An increase in heart rate		
Polycythemia		
Loss of body fluids		
Sympathetic stimulation of arterioles		
Vasodilation		
Production of angiotensin		
Slow heart beat		

Circulatory Pathways (Objectives 10-14)

1. Fill in the blanks with the correct blood vessels and valves to complete the pathway of pulmonary circulation.

   Right atrium → ________________ → Right ventricle → ______________ →

   ______________________ → ______________________ →Capillaries of lungs →

   ______________________ → Left atrium

2. Identify the arteries at the base of the brain as indicated on the following diagram.

   A. ______________________

   B. ______________________

   C. ______________________

   D. ______________________

   E. ______________________

   F. ______________________

   G. ______________________

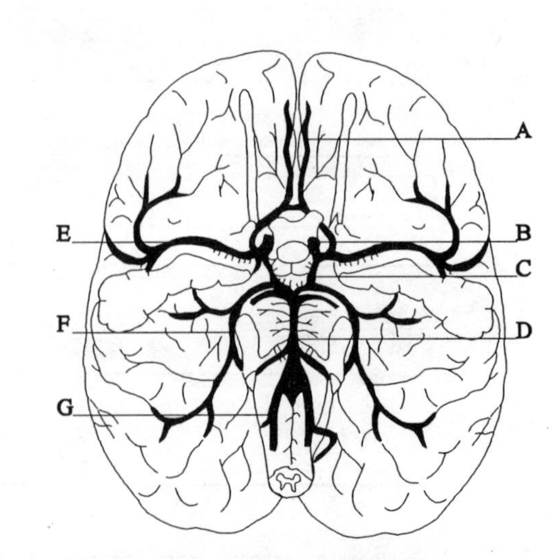

3. In the space at the left, write the name of the artery that is described.

______________________	Branches from the ascending aorta
______________________	First, or most anterior, branch of aortic arch
______________________	Branch of the aortic arch to the left arm
______________________	Paired vessels that go to the brain (2)
______________________	
______________________	Middle branch of the aortic arch

4. Name the major arteries described by each phrase.

______________________	Located in the arm
______________________	Located on lateral side of forearm
______________________	Located on medial side of forearm
______________________	Branch of celiac that supplies the liver
______________________	Branch of celiac that supplies stomach
______________________	Supplies pancreas and spleen
______________________	Supplies small intestine and part of the colon
______________________	Paired vessels to the kidneys
______________________	Supplies descending colon and rectum
______________________	Formed by bifurcation of aorta
______________________	Enters pelvis and supplies urinary bladder
______________________	Located in the thigh region
______________________	Located in the posterior knee region
______________________	Supplies the posterior portion of the leg
______________________	Supplies the anterior portion of the leg

5. Name the major veins described by each phrase.

______________________	Two large veins that enter the right atrium
______________________	
______________________	Join to form the superior vena cava
______________________	Major vein from the brain
______________________	Superficial vein on medial side of arm
______________________	Superficial vein on lateral side of arm
______________________	Vein at the elbow used for drawing blood
______________________	Deep vein in the arm
______________________	Drains muscle tissue of abdominal wall
______________________	Vein from liver that enters the IVC
______________________	Vein that carries nutrient-rich blood to liver

______________________ Two veins that form the hepatic portal vein

____

______________________ Longest vein in the body

______________________ Join to form the inferior vena cava

______________________ Drains blood from pelvic viscera

______________________ Large deep vein in the thigh

______________________ Drains dorsal foot and anterior leg muscles

______________________ Drains posterior leg muscles

6. Name the feature of fetal circulation described by each phrase.

______________________ Transports blood from fetus to placenta

______________________ Transports blood from placenta to fetus

______________________ Bypasses the liver

______________________ Opening in interatrial septum

______________________ Shunt from pulmonary trunk to aorta

______________________ Structure where gaseous exchange occurs

7. Follow the pathway for the flow of blood from the left ventricle to the right middle cerebral artery by filling in the blanks.

   Left ventricle → aortic semilunar valve → ______________________ → aortic arch →

   ______________________ → ______________________ → ______________________

   → right middle cerebral artery

8. Follow the pathway for the flow of blood from the left gonad (ovary or testicle) to the lateral side of the left forearm by filling in the blanks.

   Left gonad → left gonadal vein → ______________________ → ______________________

   → ______________________ → ______________________ valve → ______________________

   → ______________________ valve → ______________________ → ______________________

   → capillaries of lungs → ______________________ → ______________________

   → ______________________ valve → ______________________ → ______________________

   valve → ______________________ → aotic arch → ______________________

   → ______________________ → ______________________ → ______________________

   → capillaries on lateral side of left forearm

9. Follow the pathway for the flow of blood from the descending aorta to the capillaries of the spleen and then to the right atrium by filling in the blanks.

   Descending aorta → ______________________ → ______________________ → capillaries of the spleen → ______________________ → ______________________ → sinusoids of the liver

   → ______________________ → ______________________ → right atrium

☞ Chapter Self-Quiz

1. Which one of the following best describes the tunica media of arteries? (a) simple squamous epithelium; (b) smooth muscle; (c) connective tissue; (d) vasa vasorum; (e) adventitia

2. Which of the following does not describe veins? (a) they have thinner walls than arteries; (b) they have valves to prevent backflow; (c) the pressure in veins is lower than in arteries; (d) veins can hold more blood than arteries; (e) veins carry blood away from the heart

3. Indicate whether each of the following will increase (I) or decrease (D) the rate of blood flow to a specific area.

   _____ Increase the blood pressure

   _____ Increase the resistance

   _____ Increase the cardiac output

   _____ Vasodilation

   _____ Increase in carbon dioxide concentration

   _____ Contraction of precapillary sphincters

   _____ Decrease in pH

4. Indicate whether each of the following refers to (A) systolic pressure; (B) diastolic pressure, or (C) pulse pressure. You may use the letters A, B, or C as indicated.

   _____ Pressure at which first Korotkoff sound is heard

   _____ Normal is about 40 mm Hg

   _____ Blood begins to flow freely through the arteries

   _____ Normally is about 120 mm Hg

   _____ Normally is about 80 mm Hg

5. Assuming that other factors remain constant, what effect will each of the following have on blood pressure?

   I = Increase D = Decrease

   _____ Increase in heart rate

   _____ Decreased amounts of antidiuretic hormone (ADH)

   _____ Polycythemia

   _____ Decrease in blood volume

   _____ Vasodilation

   _____ Decreased elasticity in arterial walls

   _____ Stimulation of baroreceptors in the aortic arch

   _____ Epinephrine from adrenal medulla

   _____ Fluid retention

   _____ Increased angiotensin

6. Match the following descriptions with the appropriate artery. There is only one correct response for each description and not all responses will be used.

A. anterior tibial
B. brachiocephalic
C. celiac
D. common iliac
E. external iliac
F. femoral
G. hepatic
H. left common carotid
I. pulmonary
J. radial
K. renal
L. splenic
M. superior mesenteric
N. ulnar
O. vertebral

_____ Carries deoxygenated blood to the lungs

_____ Branch from the aortic arch that supplies the right arm, head, neck

_____ Formed from the bifurcation of the abdominal aorta

_____ Branch of the subclavian artery that supplies blood to the brain

_____ Artery on the medial side of the forearm

_____ Branch of the abdominal aorta that supplies blood to the liver and spleen

_____ Branch of the abdominal aorta that supplies blood to most of the small intestine

_____ Large artery in the thigh

_____ Artery that supplies blood to the anterior portion of the leg

_____ Artery that supplies blood to the kidneys

7. Match the following descriptions with the appropriate vein. There is only one correct response for each description and not all responses will be used.

A. basilic
B. brachial
C. brachiocephalic
D. cephalic
E. common iliac
F. femoral
G. great saphenous
H. hepatic
I. hepatic portal
J. inferior vena cava
K. internal jugular
L. popliteal
M. radial
N. superior vena cava
O. ulnar
P. umbilical

_____ Receives blood from the venous sinuses of the brain

_____ Carries blood from the placenta to the fetal liver

_____ Receives blood from the head and upper extremities; empties into the right atrium

_____ Deep vein in the arm

_____ Single vein in the posterior knee region

_____ Superficial vein on the lateral side of the forearm and arm

_____ Drains blood from the liver into the inferior vena cava

_____ Superficial vein of the thigh and leg; drains into the femoral vein

_____ Vessel between the superior mesenteric vein and liver

_____ Veins that join to form the inferior vena cava

☞ Terminology Exercises

WORD PART	MEANING	WORD PART	MEANING
angi-	vessel	embol-	stopper, wedge
arter-	artery	isch-	deficiency
ather-	yellow fatty plaque	phleb-	vein
brachi-	arm	scler-	hard
carotid	put to sleep	sten-	narrowing
cephal-	head	-tripsy	crushing
edem-	to swell	vas-	vessel

Use word parts given above or in previous chapters to form words that have the following definitions.

_______________________________ Pertaining to arm and head

_______________________________ A condition of arterial hardening

_______________________________ Inflammation of a vein

_______________________________ Yellow fatty tumor

_______________________________ A condition of stoppage

Using the definitions of word parts given above or in previous chapters, define the following words.

Angiology _______________________________

Arteritis _______________________________

Phlebotomy _______________________________

Edema _______________________________

Angiostenosis _______________________________

Match each of the following definitions with the correct word.

__________ Crushing a blood vessel to stop hemorrhage — A. angioma

__________ Surgical repair of a blood vessel — B. angiopathy

__________ Disease of a blood vessel — C. angioplasty

__________ Surgical excision of an artery — D. arterectomy

__________ Tumor of a blood vessel — E. vasotripsy

☞ Fun and Games

Find your way through this maze of letters by drawing a continuous line that traces the pathway of blood through the heart and pulmonary circulation. Start with the R at the arrow, draw a line to an adjacent I, then a G, and so on to complete right atrium. Continue through the heart and pulmonary circulation, including chambers, valves, vessels, and lungs, until you finish with the aorta. The consecutive letters of the words are always adjacent but may be in any direction. Do not use any letter more than one time and do not cross a line you have already drawn.

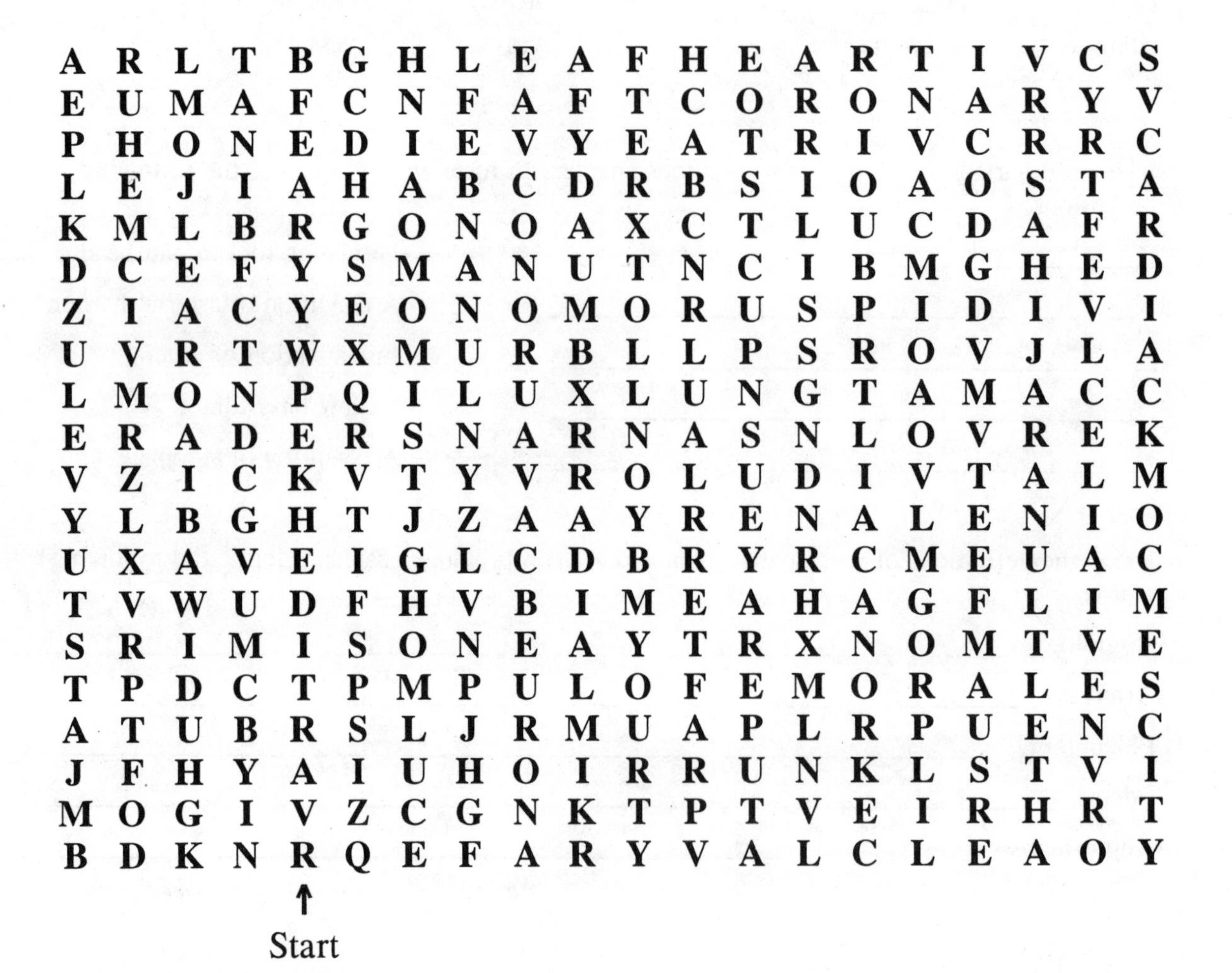

14 Lymphatic System and Body Defense

☞ Chapter Outline/Objectives

Functions of the Lymphatic System

1. State three functions of the lymphatic system.

Components of the Lymphatic System

2. List the components of the lymphatic system.
3. Describe the origin of lymph and what happens to it.
4. Name the two large lymphatic ducts and identify the region of the body each one drains.
5. State three actions that provide pressure gradients for fluid flow in lymphatic vessels.
6. Describe the structure, function, and distribution of lymph nodes.
7. Identify three groups of tonsils and describe their location and function.
8. Describe the location and structure of the spleen and discuss its functions.
9. Locate the thymus and describe its role in the production of lymphocytes.

Resistance to Disease

10. Define the terms pathogen, resistance, and susceptibility.
11. List four nonspecific mechanisms that provide resistance to disease.
12. List three types of barriers against microbial invasion.
13. Describe the actions of complement and interferon.
14. Name the two principal phagocytic cells that function in defense against disease.
15. List four signs and symptoms of inflammation and briefly describe the process of inflammation.
16. Describe three responses present in systemic inflammation that are not present in localized inflammation.
17. State the two characteristics of specific defense mechanisms and identify the two principal cells involved in specific resistance.
18. Define the term antigen.
19. Describe the development of T-lymphocytes and B-lymphocytes and state the quantity of each type.
20. Briefly describe the mechanism of cell-mediated immunity and list four subgroups of T-cells.
21. Briefly describe the mechanism of antibody-mediated immunity and list two subgroups of B-cells.
22. Distinguish between the primary response and the secondary response to a pathogen.
23. Define the term immunoglobulin.
24. List five classes of immunoglobulins and state the role each has in immunity.
25. Distinguish between active and passive immunity.
26. Distinguish between natural and artificial immunity.
27. Give examples of active natural immunity, active artificial immunity, passive natural immunity, and passive artificial immunity.

☞ Learning Exercises

Functions of the Lymphatic System (Objective 1)

1. Three functions of the lymphatic system are to:

   a.

   b.

   c.

Components of the Lymphatic System (Objectives 2-9)

1. Name the three major components o the lymphatic system.

   a.

   b.

   c.

2. In the spaces on the left, write the answers that match the statements about lymph and lymphatic vessels.

____________________	Fluid within the tissue spaces
____________________	Source of fluid in tissue spaces
____________________	Fluid within lymphatic vessels
____________________	Source of fluid within lymphatic vessels
____________________	Smallest lymphatic vessels
____________________	Largest lymphatic vessels
____________________	Drains the upper right quadrant of the body
____________________	Collects lymph from 3/4 of the body
____________________	Structure at beginning of thoracic duct
____________________	Returns lymph to right subclavian vein
____________________	Returns lymph to left subclavian vein
____________________	Prevent backflow of lymph within vessels
____________________	Three actions that move lymph in vessels
____________________	
____________________	

3. In the spaces on the left, write the answers that match the statements about the organs of the lymphatic system.

______________________	Characteristic cells in lymphatic organs
______________________	Located along lymphatic pathways
______________________	Located near the mouth, nose, and throat
______________________	Located posterior to the stomach
______________________	Located near the heart
______________________	Filters lymph
______________________	Filters blood
______________________	Protect against pathogens that enter via nose
______________________	Functions in maturation of T-lymphocytes
______________________	Reservoir for blood
______________________	Called adenoids when enlarged
______________________	Carry lymph into a lymph node
______________________	Hormone from the thymus
______________________	Consists of red pulp and white pulp

Resistance to Disease (Objectives 10-27)

1. In the spaces on the left, write the words that are defined on the right.

______________________	Disease-producing organisms
______________________	Ability to counteract harmful agents
______________________	Lack of resistance
______________________	Mechanisms against all harmful agents
______________________	Mechanisms against selected harmful agents
______________________	Specific resistance

2. Systemic inflammation is a medical crisis. List three characteristics of systemic inflammation.

 a.

 b.

 c.

3. Match each of the following processes, agents, and components with the nonspecific defense mechanism of which it is a part.

 A. Barrier to entry
 B. Phagocytosis
 C. Protective chemical action
 D. Inflammatory reaction

 _______ Unbroken skin and mucous membranes

 _______ Activates phagocytosis and inflammation

 _______ Blocks replication of viruses

 _______ Cilia action in respiratory tract

 _______ Interferon

 _______ Accompanied by swelling, heat, and redness

 _______ Ingestion and destruction of solid particles by certain cells

 _______ Action of complement

 _______ Macrophages and neutrophils

 _______ Aimed at localizing damage and destroying its source

4. In the spaces on the left, write the answers that match the statements about specific defense mechanisms.

 ______________________________ Two characteristics of immunity

 ____

 ______________________________ Two primary cell types involved in immunity

 ____

 ______________________________ Molecules that trigger immune responses

_______________ Lymphocytes that mature in the thymus

_______________ 30% of circulating lymphocytes

_______________ Responsible for cell-mediated immunity

_______________ Mature someplace other than in thymus

_______________ Responsible for humoral immunity

_______________ 70% of circulating lymphocytes

_______________ Produce antibodies

_______________ Another term for antibodies

_______________ Initial action against antigens

_______________ Four subgroups of T-cells

____

____

____

_______________ Two clones of B-cells

____

5. Write IgG, IgA, IgM, IgD, or IgE on the line preceding each phrase to identify the class of antibodies it describes.

_______ Most numerous antibody

_______ Responsible for allergic reactions

_______ Found in breast milk, saliva, mucus, and tears

_______ Responsible for ABO transfusion reactions

_______ Major antibody in immune responses

_______ Binds to mast cells and causes release of histamine

_______ Crosses placenta to provide immunity for newborn

_______ Causes agglutination of antigens

6. Match each of the following descriptions with the type of immunity it characterizes.

A. Active natural
B. Active artificial
C. Passive natural
D. Passive artificial

_______ Antiserum is injected into an individual

_______ A person contracts a disease, recovers, and is immune to that disease

_______ Antibodies are transferred from one person to another through natural processes

_______ Antigens deliberately introduced into a person to stimulate immunity

_______ Memory cells produced after contracting a disease such as chickenpox

_______ Antibodies injected into an individual after a poisonous snakebite

_______ IgA antibodies in mother's milk may provide some immunity for infant

_______ Mumps, diphtheria, whooping cough, and tetanus vaccines

☞ Chapter Self-Quiz

1. The lymphatic vessel that drains lymph from the entire body except the upper right quadrant is the (a) right lymphatic duct; (b) lacteal; (c) cisterna chyli; (d) thoracic duct

2. The primary function of lymph nodes is to (a) filter and cleanse the blood; (b) produce lymphocytes; (c) filter and cleanse lymph; (d) act as a reservoir for lymph

3. Which of the following is not a function of the spleen? (a) filter and cleanse blood; (b) produce lymphocytes; (c) act as a reservoir for blood; (d) filter and cleanse lymph

4. T-lymphocytes differentiate in the ______________________________.

5. Match each of the following with the nonspecific defense mechanism with which it is most closely associated.

   A. Barriers
   B. Chemicals
   C. Phagocytosis
   D. Inflammation

   _____ Intact skin

   _____ Interferon and complement

   _____ Lysozymes in tears

   _____ Swelling and pain

   _____ Neutrophils and macrophages

   _____ Fluid flow (urine)

6. The primary cells involved in specific resistance are ______________________________

   and ______________________________.

7. Indicate whether each of the following refers to T-cells, B-cells, or both T- and B- cells.

T = T-cells   B = B-cells   TB = both

___________ Memory cells

___________ Cell-mediated immunity

___________ Plasma cells

___________ Humoral immunity

___________ Antibodies

___________ Clones of helper and killer cells

___________ Macrophage presents antigen

___________ Immunoglobulins

8. The type of immunity acquired by obtaining a vaccination that consists of a weakened pathogen is (a) active natural immunity; (b) active artificial immunity; (c) passive natural immunity; (d) passive artificial immunity.

☞ Terminology Exercises

WORD PART	MEANING
aden-	gland
-ectomy	surgical removal
immun-	protection
lymph-	lymph
-megaly	large
-pexy	fixation
splen-	spleen
tox-	poison
thym-	thymus

Use word parts given above or in previous chapters to form words that have the following definitions.

______________________ Surgical excision of a tonsil

______________________ Fixation of a movable spleen

______________________ Stopping flow of lymph

______________________ Study of body protection

______________________ Inflammation of the thymus

Using the definitions of word parts given above or in previous chapters, define the following words.

Thymoma______________________

Splenomegaly ______________________

Lymphangiology ______________________

Lymphadenitis ______________________

Lymphocytopenia ______________________

Match each of the following definitions with the correct word.

________	Poison substance	A.	lymphadenotomy
________	Surgical removal of the thymus	B.	lymphopoiesis
________	Incision into a lymph gland	C.	splenemia
________	Condition of spleen congested with blood	D.	thymectomy
________	Formation of lymph	E.	toxin

☞ Fun and Games

Fill in the blanks by answering the clues at the bottom of the page.

1. _ _ _ _ I _ _
2. _ _ _ M _ _
3. _ _ M _ _ _ _ _ _ _
4. _ _ _ U _ _ _ _ _ _ _ _ _ _
5. _ N _ _ _ _ _ _ _ _ _
6. _ _ _ _ O _ _ _ _ _
7. _ _ _ _ _ _ L _ _ _ _
8. _ _ _ _ _ O _ _ _ _
9. _ _ _ _ G _ _
10. _ Y _ _ _ _ _ _
11. _ _ A _ _ _   _ _ _ _
12. _ N _ _ _ _   _ _ _ _
13. _ _ T _ _ _ _ _ _ _
14. _ _ _ _ I _ _ _ _ _ _
15. _ _ B _ _
16. _ _ _ _ O _ _ _ _ _ _ _
17. _ _ _ _ _ _ D _ _ _ _ _ _
18. _ _ _ I _ _
19. _ _ _ _ _ E _ _
20. _ _ S _ _ _ _ _ _ _ _ _ _ _

CLUES

1. Lymphoid tissue in pharynx
2. Processes certain lymphocytes
3. WBC involved in immunity
4. Another name for antibodies
5. Protein that stimulates T-cells
6. Large phagocytic cell
7. Exaggerated hypersensitivity response
8. Small phagocytic cell
9. Non-self molecule
10. Enzyme in tears and saliva
11. Produces antibodies
12. Mechanical barrier against pathogens
13. Inhibits viral replication
14. Injection of inactivated pathogens
15. Sign of inflammation
16. Cellular ingestion of pathogens
17. Inflammation of lymph glands
18. Type of immunity
19. Cause fever
20. Opposite of resistance

15 Respiratory System

☞ Chapter Outline/Objectives

Functions and Overview of Respiration

1. Define five activities of the respiratory process.

Ventilation

2. Distinguish between the upper respiratory tract and the lower respiratory tract.
3. Label a diagram showing the features of the nasal cavities.
4. List three functions of the nasal cavities.
5. Name the three regions of the pharynx, state where each is located, and identify the features unique to each region.
6. Describe the location of the larynx and identify the three largest cartilages in its framework.
7. Describe the framework of the trachea and identify the type of tissue that forms the lining.
8. List in sequence the branches of the bronchial tree beginning with the bifurcation of the trachea and ending with the alveolar ducts.
9. Name the type of tissue that forms the alveoli.
10. Compare the right lung and left lung by shape and number of lobes.
11. Define parietal pleura, visceral pleura, and pleural cavity.
12. Name and define three pressures involved in pulmonary ventilation.
13. Describe the role of the diaphragm in inspiration and expiration.
14. List the sequence of events that results in inspiration and expiration.
15. Explain the importance of surfactant in maintaining inflated alveoli.
16. Define four respiratory volumes and four respiratory capacities and state their average normal values.
17. Describe four factors that may influence lung volumes and capacities.

Basic Gas Laws and Respiration

18. State Dalton's law of partial pressures and Henry's gas law.
19. List the six layers of the respiratory membrane through which gases diffuse in external respiration.
20. Discuss three factors that affect the rate at which external respiration occurs.
21. Distinguish between external respiration and internal respiration.

Transport of Gases

22. Describe two ways in which oxygen is transported in the blood.
23. Discuss three mechanisms of carbon dioxide transport in the blood.

Regulation of Respiration

24. Name two regions in the brain that make up the respiratory center and two nerves that carry impulses from the center.
25. Describe the role of chemoreceptors, stretch receptors, higher brain centers, and temperature in regulating breathing.

☞ Learning Exercises

Functions and Overview of Respiration (Objective 1)

1. In the spaces on the left, write the terms that match the phrases about the sequence of events in respiration.

_______________ Exchange of air between atmosphere and lungs

_______________ Exchange of gases between lungs and blood

_______________ Function of blood in respiration

_______________ Exchange of gases between blood and tissue cells

_______________ Utilization of oxygen in the cells

Ventilation (Objectives 2-17)

1. Identify structures related to the nasal cavity by matching the letters from the diagram with the correct terms.

_______ Ethmoidal sinus

_______ Frontal sinus

_______ Hard palate

_______ Nasal cavity

_______ Opening for auditory tube

_______ Oral cavity

_______ Pharynx

_______ Soft palate

_______ Sphenoidal sinus

E
F
A
G
H
B
C
I
D

2. Use the letter U to designate the regions of the upper respiratory tract and the letter L to designate the regions of the lower respiratory tract.

_______ Bronchi

_______ Larynx

_______ Lungs

_______ Nose

_______ Pharynx

_______ Trachea

3. In the spaces on the left, write the answers that match the phrases about the respiratory tract.

_______________ Three functions of the nasal cavity

____

____

________________ Region of pharynx posterior to nasal cavity

________________ Location of pharyngeal tonsils

________________ Pharynx from uvula to hyoid bone

________________ Lowest region of pharynx

________________ Cartilage that forms the Adam's apple

________________ Most inferior cartilage of larynx

________________ Prevents food from entering larynx

________________ Opening between the true vocal cords

________________ Passage commonly called the windpipe

________________ Ridge of cartilage at bifurcation of trachea

________________ Tissue that forms lining of the trachea

________________ Tissue that forms the alveoli

4. Identify the indicated structures on the diagram by matching the letters with the correct terms on the left.

_______ Carina

_______ Cricoid cartilage

_______ Primary bronchus

_______ Secondary bronchus

_______ Thyroid cartilage

_______ Trachea

5. Arrange the following air passages in the correct sequence from largest to smallest by placing the numbers 1-7 in the spaces before the names. Use number 1 for the largest and number 7 for the smallest.

_______ Alveolar ducts

_______ Alveoli

_______ Lobar bronchi

_______ Primary bronchi

_______ Respiratory bronchioles

_______ Segmental bronchi

_______ Terminal bronchioles

6. Place an R in the space preceding the phrase if it refers to the right lung. Place an L in the space if it refers to the left lung. If the phrase applies to both lungs, place a B in the space.

_______	Cardiac notch	______	Two lobes
_______	Shorter and wider	______	Divided into lobules
_______	Rests on the diaphragm	______	Has two fissures
_______	Enclosed by the pleura	______	Anchored at the root or hilum

7. In the spaces on the left, write the answers that match the phrases about pulmonary ventilation.

______________________________	Name of pressure between layers of pleurae
______________________________	Name of pressure outside the body
______________________________	Name of pressure within the alveoli
______________________________	Pressure that is normally less than other two
______________________________	Primary muscle involved in quiet breathing
______________________________	Muscles used in forced expiration
______________________________	Reduces surface tension within alveoli
______________________________	Instrument to measure respiratory volumes
______________________________	Highest pressure during expiration
______________________________	Highest pressure during inspiration

8. Write I in the space if the event occurs during inspiration and write E if it occurs during expiration.

_______ Diaphragm contracts

_______ Intrapulmonary pressure exceeds atmospheric pressure

_______ External intercostal muscles may contract

_______ Atmospheric pressure is greater than intrapulmonary pressure

_______ Lung volume increases

_______ Diaphragm relaxes

_______ Internal intercostal muscles may contract

_______ Air flows into the lungs

_______ Elastic recoil decreases size of alveoli

9. Match the lung volumes and capacities with their definitions.

TV = Tidal volume
IRV = Inspiratory reserve volume
ERV = Expiratory reserve volume
RV = Residual volume
VC = Vital capacity
IC = Inspiratory capacity
FRC = Functional residual capacity
TLC = Total lung capacity

_______ Maximum amount of air that can be inhaled
_______ Amount of air inhaled and exhaled in quiet breathing cycle
_______ Amount of air in lungs after quiet expiration
_______ Equals TV + IRV + ERV
_______ Maximum amount of air that can be inhaled after a tidal inspiration
_______ Amount of air in lungs after maximum inspiration
_______ Maximum amount of air that can be forcefully exhaled after tidal expiration
_______ Equals RV + ERV
_______ Amount of air in lungs after maximum expiration
_______ Equals RV + ERV + TV + IRV

10. Complete the following table by writing in the correct values in the blank spaces.

TLC	VC	TV	ERV	IRV	RV
		500 ml	1,000 ml	3,000 ml	1,200 ml
6,000 ml	5,000 ml		900 ml	3,500 ml	
	4,700 ml	600 ml	800 ml		1,100 ml
5,900 ml	4,600 ml	400 ml		3,000 ml	

11. If the individual type on the left tends to have greater lung volumes than the type on the right, write > (greater than) on the blank in the center. If the left type tends to be less than the right, write < (less than).

Young adults _______ Senior citizens
Females _______ Males
Short people _______ Tall people
Normal weight people _______ Obese people
Healthy people _______ People with muscular disease
Good physical condition _______ Poor physical condition

Basic Gas Laws and Respiration (Objectives 18-21)

1. Atmospheric air is a mixture of oxygen, nitrogen, and carbon dioxide. According to Dalton's law of partial pressures, if the atmospheric pressure is 750 mm Hg and the oxygen content is 21%, what is the pressure due to oxygen?

2. According to Henry's law, what two factors determine how much of each gas in a mixture will dissolve in a liquid?

3. Complete the following statements that define external and internal respiration.

   External respiration is the exchange of gases between the ____________________

   ____

   Internal respiration is the exchange of gases between the ____________________

   ____

4. List, in sequence, the six layers of the respiratory membrane through which oxygen must pass to diffuse from the alveolus into the blood capillary.

   a. d.

   b. e.

   c. f.

5. Indicate whether each of the following will increase or decrease the rate of gaseous exchange across the respiratory membrane. Use I for increase and D for decrease.

   _______ Decreasing the surface area of the respiratory membrane

   _______ Fluid accumulation in the alveoli due to pulmonary edema

   _______ Increasing tidal volume

   _______ Decreasing breathing rate

Transport of Gases (Objectives 22 and 23)

1. List two ways in which oxygen is transported in the blood.

   a.

   b.

2. List three ways in which carbon dioxide is transported in the blood.

   a.

   b.

   c.

3. In the spaces on the left, write the answers that match the phrases about oxygen and carbon dioxide transport.

____________________	Compound in which most oxygen is transported
____________________	Method of most carbon dioxide transport
____________________	Combination of carbon dioxide and hemoglobin
____________________	Combination of water and carbon dioxide
____________________	Enzyme that speeds up the reaction between water and carbon dioxide

4. Place a check (✓) in front of the factors that favor unloading of oxygen in the tissues.

   ______ Increased partial pressure of oxygen

   ______ Increased partial pressure of carbon dioxide

   ______ Increased hydrogen ion concentration

   ______ Increased temperature

   ______ Increased pH

   ______ Increased cellular metabolism

5. Place an E before the phrases that refer to external respiration and place an I before the phrases that refer to internal respiration.

   ______ Oxygen diffuses into the blood

   ______ Oxygen diffuses out of the blood

   ______ Carbon dioxide diffuses into the blood

   ______ Carbon dioxide diffuses out of the blood

   ______ Occurs in the alveolus

   ______ Occurs in the body tissues

   ______ Bicarbonate ion is formed

   ______ Bicarbonate ions release carbon dioxide

   ______ Oxyhemoglobin is formed

   ______ Oxyhemoglobin dissociates and releases oxygen

Regulation of Respiration (Objectives 24 and 25)

1. The respiratory center includes neurons in the ____________ and ______________.

2. The inspiratory areas of the respiratory center send impulses along the ___________ nerve to the diaphragm and along the _______________________________ nerves to the external intercostal muscles.

3. Place a T before each true statement and an F before each false statement about factors that influence the rate and depth of breathing.

_______ Chemoreceptors in the medulla oblongata are sensitive to changes in oxygen levels.

_______ Increases in blood CO_2 levels increase the rate and depth of breathing.

_______ Increases in hydrogen ion concentrations increase the rate and depth of breathing.

_______ Chemoreceptors in the medulla oblongata are sensitive to changes in carbon dioxide and hydrogen ion concentrations.

_______ A decrease in oxygen levels is usually a strong stimulus for breathing.

_______ Decreased oxygen levels usually make the respiratory center more sensitive to carbon dioxide changes.

_______ Peripheral chemoreceptors are located in the aortic and carotid bodies.

_______ The Hering-Breuer reflex prevents overinflation of the lungs.

_______ The Hering-Breuer reflex is a response to stretch receptors in the lungs.

_______ Higher brain centers may permanently override the respiratory center.

_______ Anxiety decreases the rate and depth of breathing.

_______ Chronic pain stimulates breathing, but sudden pain may cause a momentary cessation of breathing.

_______ Decreasing body temperature increases the breathing rate.

_______ The primary stimulus for breathing is decreased carbon dioxide levels in the respiratory center.

☞ Chapter Self-Quiz

1. The pathway of inhaled air is
   (a) nasal cavity, trachea, larynx, bronchi, pharynx, alveoli
   (b) nasal cavity, pharynx, larynx, trachea, bronchi, alveoli
   (c) nasal cavity, larynx, pharynx, trachea, bronchi, alveoli
   (d) nasal cavity, pharynx, trachea, bronchi, larynx, alveoli
   (e) nasal cavity, pharynx, trachea, larynx, bronchi, alveoli.

2. The following pairs of terms indicate a region of the pharynx and a structure or opening located in that region. Which pair is mismatched?
   (a) nasopharynx - openings for auditory tubes
   (b) oropharynx - fauces
   (c) oropharynx - palatine tonsils
   (d) laryngopharynx - pharyngeal tonsils
   (e) nasopharynx - adenoids

3. The following statements are about the lungs and pleura. Place a T before the true statements and an F before the false statements.

   _____ The left lung has three lobes and is longer than the right lung.

   _____ The pleural cavity is between the visceral pleura and the alveoli.

   _____ The heart makes an indentation, called the cardiac notch, in the left lung.

   _____ The lungs are divided into bronchopulmonary segments, each with its own segmental bronchus.

4. Given that atmospheric pressure is 760 mm Hg, intraalveolar pressure is 763 mm Hg, and intrapleural pressure is 756 mm Hg. These pressures indicate the
   (a) inspiration phase of ventilation
   (b) expiration phase of ventilation
   (c) period after expiration but before the next inspiration

5. Given that for a particular patient the tidal volume = 450 ml, inspiratory reserve volume = 2800 ml, expiratory reserve volume = 1050 ml, and residual volume = 1200 ml, what is this patient's vital capacity?
   (a) 1500 ml
   (b) 3250 ml
   (c) 5500 ml
   (d) 4300 ml
   (e) 3850 ml

6. Which of the following is <u>not</u> true about external respiration?
   (a) carbon dioxide enters the blood and oxygen enters the alveoli
   (b) gases diffuse across the respiratory membrane in the lungs
   (c) an accumulation of fluid in the alveoli decreases the diffusion rate
   (d) surface area of the alveoli affects the diffusion rate

7. The affinity of hemoglobin and oxygen decreases when
   (a) hydrogen ion concentration decreases
   (b) carbon dioxide levels increase
   (c) temperature decreases
   (d) oxygen levels increase

8. More than 2/3 of the carbon dioxide transported in the blood is
   (a) dissolved in the plasma
   (b) bound to hemoglobin
   (c) attached to carbonic anhydrase
   (d) in the form of bicarbonate ions

9. The respiratory center in the brain includes neurons in the
   (a) cerebrum and cerebellum
   (b) thalamus and pons
   (c) midbrain and medulla oblongata
   (d) medulla oblongata and pons

10. Chemoreceptors in the central nervous system stimulate inspiration when they detect
    (a) low oxygen levels
    (b) low hydrogen ion and high carbon dioxide levels
    (c) increased hydrogen ion and carbon dioxide levels
    (d) increased hydrogen ion and decreased carbon dioxide levels

☞ Terminology Exercises

WORD PART	MEANING	WORD PART	MEANING
-a	lack of	-pnea	breathing
alveol-	tiny cavity	pneum-	lung, air
anthrac-	coal	-ptysis	spitting
bronchi-	bronchi	pulmon-	lung
-coni-	dust	-rrhea	flow or discharge
cric-	ring	rhin-	nose
dys-	difficult	spir-	breath
-ectasis	dilation	thyr-	shield
eu-	good	-tion	act of, process of
phon-	voice	ventilat-	to fan or blow

Use word parts given above or in previous chapters to form words that have the following definitions.

_______________________________ Lack of breathing

_______________________________ Resembles a ring

_______________________________ Surgical repair of the nose

_______________________________ Dilation of bronchi

_______________________________ Rapid breathing

Using the definitions of word parts given above or in previous chapters, define the following words.

Hemoptysis _______________________________

Thyroid _______________________________

Rhinitis _______________________________

Dyspnea _______________________________

Pneumoconiosis _______________________________

Match each of the following definitions with the correct word.

__________	Discharge from the nose	A. alveoli
__________	Presence of tiny cavities	B. anthracosis
__________	Process of voice production	C. phonation
__________	Condition caused by inhaling coal dust	D. pulmonectomy
__________	Surgical removal of a lung	E. rhinorrhea

☞ Fun and Games

Each of the following clues is preceded by a plus (+) or minus (−) sign. Answer each clue. The number in parentheses indicates the number of letters in the answer to the clue. If there is a plus (+) preceding the clue, add the letters of your answer to the letters from previous answers. If there is a minus (−) preceding the clue, remove the letters of your answer from the previous letters. When you finish, there should be one letter remaining.

Example:

\+ 1. A feline (3) CAT
Letters: CAT

\+ 2. A rodent (3) RAT
Letters: CATRAT

− 3. A type of wagon (4) CART
Letters: AT

− 4. First letter of the alphabet (1) A
Letters: T

The final letter remaining is T. T is for TIME. It is TIME now for Fun and Games.

CLUES

\+ 1. Throat (7)
Letters:

\+ 2. Membrane around lungs (6)
Letters:

− 3. Voice box (6)
Letters:

\+ 4. Reduces surface tension (10)
Letters:

\+ 5. Root for "mouth" or "opening" (2)
Letters:

− 6. Cessation of breathing (5)
Letters:

\+ 7. Nostrils (5)
Letters:

− 8. Opening from mouth into pharynx (6)
Letters:

− 9. Prefix for "lacking" or "without" (1)
Letters:

\+ 10. Bifurcation of trachea (6)
Letters:

CLUES

\+ 11. Air sac in lungs (8)
Letters:

− 12. Swiss mountain (3)
Letters:

− 13. Continues after larynx (7)
Letters:

\+ 14. Lung collapse (11)
Letters:

\+ 15. Major respiratory muscle (9)
Letters:

− 16. What the telephone did (4)
Letters:

− 17. Whooping cough (9)
Letters:

− 18. Air breathed in and out (11)
Letters

− 19. Coal dust in lungs (11)
Letters:

FINAL LETTER:

Hint: The final letter is the first letter of the topic for this chapter.

16 Digestive System

☞ Chapter Objectives

Introduction

1. List the components of the digestive tract and the accessory organs.

Functions of the Digestive System

2. List six functions of the digestive system.

General Structure of the Digestive Tract

3. Describe the general structure of the four layers, or tunics, in the digestive tract wall.

Components of the Digestive Tract

4. List and describe the boundaries of the oral cavity.
5. Describe the structure and functions of the tongue.
6. Distinguish between the primary and secondary teeth.
7. Show the relationship between tooth shape and function.
8. Identify the features of a "typical" tooth.
9. Name and describe the location of the three major types of salivary glands.
10. List the components and describe the functions of saliva.
11. Identify and describe the features of the pharynx.
12. Describe the location and features of the esophagus.
13. Identify the features of the stomach.
14. Describe how the structure of the stomach wall differs from the generalized structure of the digestive tract.
15. Name the types of cells found in gastric glands and tell what is secreted by each type.
16. Describe the events in each of the three phases in the regulation of gastric secretion.
17. Discuss two factors that affect the rate at which the stomach contents enter the small intestine.
18. Describe three features of the small intestine that increase the surface area for absorption.
19. Distinguish between the three regions of the small intestine with respect to location, length, and features.
20. Name five digestive enzymes and two hormones that are produced in the small intestine and indicate the function of each one.
21. Describe how chyme affects intestinal secretions.
22. Describe how the structure of the wall of the large intestine differs from the generalized structure of the digestive tract.
23. List the regions of the large intestine and state the location of each region.
24. List two functions of the large intestine.

Accessory Organs of Digestion

25. Identify the external features of the liver.
26. Draw and label a diagrammatic representation of a liver lobule.
27. Distinguish between the two sources of blood for the liver and trace the flow of blood through the liver.
28. List 10 functions of the liver.
29. Describe the role of bile in digestion and as an excretory agent.

30. Describe the location and function of the gallbladder.
31. Describe the location of the pancreas.
32. Distinguish between the exocrine and endocrine portions of the pancreas.
33. List four pancreatic enzymes and explain the function of each one.
34. Name two hormones that regulate pancreatic secretions and distinguish between the functions of each one.

Chemical Digestion

35. Summarize carbohydrate digestion by writing an equation that shows the intermediate and final products and the enzymes that facilitate the digestive process.
36. Summarize protein digestion by writing an equation that shows the intermediate and final products and the enzymes that facilitate the digestive process.
37. Summarize lipid digestion by writing an equation that shows the intermediate and final products and the factors that facilitate the digestive process.
38. Compare the absorption of simple sugars and amino acids with that of lipid related molecules.

☞ Learning Exercises

Introduction (Objective 1)

1. Identify the parts of the digestive system by matching the letters from the figure on the right with the terms on the left. Place an asterisk (*) by the accessory organs.

_______ Esophagus

_______ Gallbladder

_______ Large intestine

_______ Liver

_______ Mouth

_______ Pancreas

_______ Pharynx

_______ Salivary gland

_______ Small intestine

_______ Stomach

Functions of the Digestive System (Objective 2)

1. Write the terms that are described by the phrases about the functions of the digestive system.

______________________	Process of breaking large food particles into smaller ones
______________________	Uses water to break down large molecules into smaller ones
______________________	Breaks down complex nonabsorbable molecules into simple usable molecules
______________________	Swallowing
______________________	Movements that propel food through the digestive tract
______________________	Process by which simple molecules from chemical digestion pass through the cell membranes into the blood
______________________	Removal of indigestible wastes through the anus

General Structure of the Digestive Tract (Objective 3)

1. Match the following descriptions with the correct layer of the digestive tract.

A. Mucosa
B. Submucosa
C. Muscular layer
D. Serosa

_______ Innermost layer of the digestive tract wall

_______ Contains blood and lymphatic vessels embedded in loose connective tissue

_______ Responsible for most movements of the digestive tract

_______ Consists of simple columnar epithelium in stomach and intestines

_______ Contains inner circular and outer longitudinal layers

_______ Contains Meissner's plexus of autonomic nerve fibers

_______ Contains the myenteric plexus of autonomic nerve fibers

Components of the Digestive Tract (Objectives 4-24)

1. Write the terms that match the phrases about the oral cavity.

______________________	Primary muscle in the cheek
______________________	Separates oral cavity from nasal cavity
______________________	Projection at posterior end of soft palate
______________________	Masses of lymphoid tissue at back of tongue
______________________	Projections on surface of tongue

______________________	Teeth with sharp edges for biting
______________________	Teeth with points for grasping and tearing
______________________	Largest salivary glands
______________________	Glands along medial surface of mandible
______________________	Enzyme found in saliva
______________________	Four functions of saliva
______________________	
______________________	
______________________	

2. Identify the parts of a tooth by matching the letters from the diagram with the correct terms.

_______ Alveolar process

_______ Apical foramen

_______ Cementum

_______ Dentin

_______ Enamel

_______ Gingiva

_______ Pulp cavity

_______ Root canal

3. Write the terms that match the phrases about the pharynx and esophagus.

______________________	Opening from oral cavity into pharynx
______________________	Most superior region of the pharynx
______________________	Region that contains pharyngeal tonsils
______________________	Region that contains palatine tonsils
______________________	Keeps food from entering nasopharynx
______________________	Keeps food from entering laryngopharynx
______________________	Tube between pharynx and stomach
______________________	Opening in diaphragm for esophagus
______________________	Sphincter between esophagus and stomach

4. Identify the regions of the stomach by matching the letters from the diagram with the correct terms. Outline the greater curvature in red and the lesser curvature in blue.

_______ Body
_______ Cardiac region
_______ Duodenum
_______ Fundus
_______ Lower esophageal sphincter
_______ Pyloric sphincter
_______ Pylorus
_______ Rugae

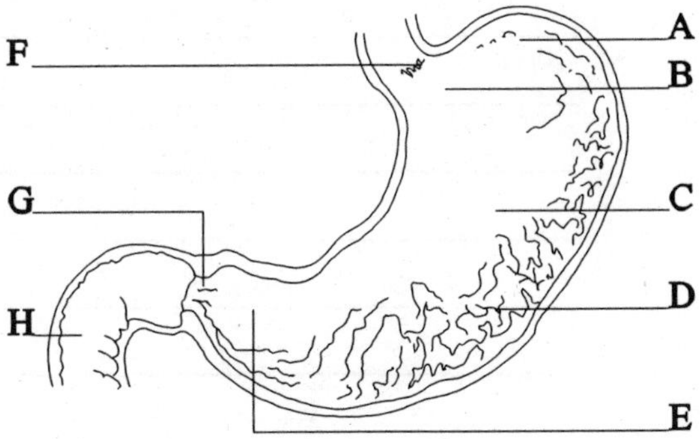

F
G
H
A
B
C
D
E

5. Write the terms that match the phrases about the stomach and its secretions.

______________________________	Acid in gastric juice
______________________________	Hormone secreted by gastric mucosa
______________________________	Enzyme in gastric juice
______________________________	Semifluid mixture of food and gastric juice
______________________________	Function of intrinsic factor
______________________________	Two factors that influence stomach emptying
______________________________	

6. Match the events in the regulation of gastric secretions with the phase in which they occur.

A. Cephalic phase B. Gastric phase C. Intestinal phase

_______ Triggered by the passage of chyme into the small intestine
_______ Begins with thoughts of food
_______ Begins when food reaches stomach
_______ Inhibits gastric secretions
_______ Involves distention of the stomach wall

7. Write the terms that match the phrases about the small intestine and its secretions.

______________________________	Region adjacent to stomach
______________________________	Three features that increase surface area
______________________________	
______________________________	
______________________________	Lymph capillary in a villus

________________ Has mucous glands in submucosa

________________ Region with most goblet cells

________________ Activates a pancreatic enzyme

________________ Stimulates release of bile from gallbladder

________________ Stimulates pancreatic release of bicarbonate

________________ Acts on neutral fats

________________ Stimulates pancreatic digestive enzymes

________________ Most important factor for regulating secretions of the small intestine

8. What are the two main functions of the large intestine?

A.

B.

9. Match the descriptive phrases with the correct terms.

A. Anus
B. Ascending colon
C. Cecum
D. Descending colon
E. Epiploic appendages
F. Haustra
G. Hepatic flexure
H. Ileocecal junction
I. Rectum
J. Sigmoid colon
K. Splenic flexure
L. Teniae coli
M. Transverse colon

_______ Where small intestine enters the large intestine

_______ Three bands of longitudinal muscle in large intestine

_______ Pieces of fat-filled connective tissue attached to colon

_______ Blind pouch that extends inferiorly from entrance of ileum

_______ Portion of large intestine on the right side

_______ Right colonic flexure

_______ Left colonic flexure

_______ Portion of large intestine between the two colonic flexures

_______ Portion of large intestine on the left side

_______ S-shaped curve across the pelvic brim

_______ Portion that follows the curvature of the sacrum

_______ Terminal opening of the digestive tract

_______ Series of pouches in the large intestine

Accessory Organs of Digestion (Objectives 25-34)

1. Write the terms that match the phrases about the liver.

_______________ Attaches liver to anterior abdominal wall

_______________ Region between ligamentum venosum and IVC

_______________ Region between ligamentum teres and gallbladder

_______________ Functional unit of liver

_______________ Venous channels between hepatocytes

_______________ Vessel that carries oxygen-rich blood to the liver

_______________ Vessel that carries nutrient-rich blood to the liver

_______________ Vessel that carries blood away from the liver

_______________ Secretory product of the liver

_______________ Principal bile pigment

_______________ Function of bile salts

_______________ Area of bile storage

_______________ Duct that attaches gallbladder to liver

_______________ Stimulates release of bile from gallbladder

_______________ Duct that enters duodenum

_______________ Three components of a portal triad

____

____

2. Write the terms that match the phrases about the pancreas.

_______________ Endocrine portion of the pancreas

_______________ Exocrine portion of the pancreas

_______________ Carries pancreatic enzymes to duodenum

_______________ Pancreatic enzyme that acts on starch

_______________ Pancreatic enzyme that acts on proteins

_______________ Breaks fats into fatty acids and monoglycerides

_______________ Two hormones that regulate pancreatic secretions

____

Chemical Digestion (Objectives 35-37)

1. Complete the following table about enzymes and their role in digestion.

Enzyme	Source	Digestive Action
	Salivary glands/ pancreas	Complex carbohydrates to disaccharides
		Maltose to glucose
Sucrase		
		Lactose to glucose and galactose
	Stomach	Proteins to polypeptides
Trypsin		Proteins to polypeptides
		Peptides to amino acids
		Fats to fatty acids and monoglycerides

2. What are the three monosaccharides that are the end products of carbohydrate digestion?

3. What are the end products of protein digestion?

4. What are the two primary end products of lipid digestion?

Absorption (Objective 38)

1. Write the terms that match the phrases about the absorption of nutrients.

_________________________ Villus vessel that absorbs amino acids and simple sugars

_________________________ Villus vessel that absorbs triglycerides

_________________________ Fatty acids coated with bile salts

_________________________ Triglycerides combined with proteins

_________________________ Mixture of lymph and digested fats

2. Write the transport process that is responsible for each of the following movements.

_________________________ Glucose from lumen of small intestine into villus cells

_________________________ Fructose from lumen of small intestine into villus cells

_________________________ Water throughout the length of the digestive tract

_________________________ Most nutrients from villus cells to capillaries and lacteals

☞ Chapter Self-Quiz

1. Which of the following terms refers to chemical digestion? (a) mastication; (b) hydrolysis; (c) peristalsis; (d) defecation; (e) deglutition

2. Starting from the inside, or lumen, the sequence of layers in the wall of the gastrointestinal tract is (a) serosa, mucosa, submucosa, muscular; (b) mucosa, submucosa, inner longitudinal muscle, outer circular muscle; (c) serosa, muscular, submucosa, mucosa; (d) mucosa, submucosa, muscular, serosa; (e) inner muscle, submucosa, mucosa, outer muscular

3. How many **more** teeth are there in a complete permanent set than in a complete deciduous set?

4. **True or False** The largest of the salivary glands is the parotid gland, which is located in the oropharynx.

5. Match each of the following descriptions with the appropriate structure. Each description has only one response. Responses are to be used only once and not all responses will be used.

A. duodenum
B. fauces
C. fundus
D. haustra
E. hepatic flexure
F. ileocecal sphincter
G. ileum
H. jejunum
I. lower esophageal sphincter
J. plicae circulares
K. pyloric sphincter
L. pylorus
M. rugae
N. splenic flexure
O. teniae coli

____ Opening from the oral cavity into the pharynx

____ Portion of the stomach that is superior to the entrance of the esophagus

____ Circular band of muscle at the exit from the stomach

____ Circular folds of mucosa in the small intestine

____ Circular band of muscle between the esophagus and stomach

____ Curve between the ascending colon and the transverse colon

____ Longitudinal folds of mucosa in the stomach

____ Longitudinal muscle bands in the large intestine

____ Shortest part of the small intestine

____ Circular band of muscle between the small intestine and large intestine

6. Match the following actions with the appropriate enzyme or hormone. Some responses may be used more than once and others may not be used at all.

A. amylase
B. cholecystokinin
C. maltase
D. secretin
E. trypsin

_____ Breaks carbohydrates into disaccharides

_____ Stimulates secretion of bile

_____ Acts on a disaccharide

_____ Stimulates secretion of bicarbonate from pancreas

_____ Breaks proteins into peptides

_____ Stimulates secretion of enzymes from the pancreas

7. Which of the following represents the correct sequence of blood flow from the gastrointestinal tract through the liver? (a) hepatic artery → sinusoids → hepatic portal vein → hepatic vein → inferior vena cava; (b) hepatic portal vein → sinusoids → central vein → hepatic vein → inferior vena cava; (c) hepatic vein → sinusoids → central vein → hepatic portal vein → inferior vena cava; (d) hepatic portal vein → hepatic vein → sinusoids → central vein → inferior vena cava; (e) hepatic vein → hepatic portal vein → central vein → sinusoids → inferior vena cava

8. What function of the liver pertains to the digestion of fats?

9. What two secretory ducts empty into the duodenum?

10. The following six items are end products of digestion. Indicate whether each is a product of carbohydrate (C), protein (P), or fat (F) digestion and whether it is absorbed by the blood capillaries (B) or lacteals (L) of the villi. Each end product should be designated by a C, P, or F **and** by a B or L.

C, P, F	B or L	
_____	_____	Glucose
_____	_____	Amino acids
_____	_____	Monoglycerides
_____	_____	Fructose
_____	_____	Fatty acids
_____	_____	Galactose

☞ Terminology Exercises

WORD PART	MEANING	WORD PART	MEANING
-algia	pain	gastr-	stomach
amyl-	starch	gingiv-	gums
-ary	pertaining to	hepat-	liver
-ase	enzyme	lingu-	tongue
bili-	bile, gall	-orexia	appetite
chole-	gall, bile	prandi-	meal
cyst-	bladder	proct-	rectum, anus
dent-	tooth	-rrhea	flow or discharge
-emesis	vomit	sial-	saliva
enter-	intestine	verm-	worm

Use word parts given above or in previous chapters to form words that have the following definitions.

________________________________ Enzyme that breaks down starch

________________________________ Lack of appetite

________________________________ Inflammation of the intestine

________________________________ Pain in the stomach

________________________________ Pertaining to below the tongue

Using the definitions of word parts given above or in previous chapters, define the following words.

Biliary ________________________________

Cholecystitis ________________________________

Dentalgia ________________________________

Hepatitis ________________________________

Sialadenitis ________________________________

Match each of the following definitions with the correct word.

__________	After a meal	A. cholecystectomy
__________	Surgical excision of the gallbladder	B. gingivitis
__________	Vomiting blood	C. hematemesis
__________	Pain in the rectum and anus	D. postprandial
__________	Inflammation of the gums	E. proctalgia

☞ Fun and Games

This is a variation of the word game "Hangman." Guess any letter for the first word and find the number that corresponds to that letter in the Letter Chart at the bottom of the page. Then find that same number above the line that divides each cell in the Position Chart on the right. If the letter you guessed appears in the word, its position is given by the number or numbers below the line that divides each cell in the Position Chart. If the letter does not appear in the word, 0 will be indicated under the line. If the letter you guessed does not appear in the word, start drawing a "stick person" on a gallows--first a head, then a body, followed by two arms and two legs. You are allotted six wrong guesses before you are "hanged." Clue: All words are from this chapter on the digestive system.

WORDS

#1 ___ ___ ___ ___ ___ ___ ___
1 2 3 4 5 6 7

#2 ___ ___ ___ ___ ___ ___ ___ ___
1 2 3 4 5 6 7 8

#3 ___ ___ ___ ___ ___ ___ ___ ___
1 2 3 4 5 6 7 8

#4 ___ ___ ___ ___ ___ ___ ___ ___
1 2 3 4 5 6 7 8

#5 ___ ___ ___ ___ ___ ___ ___ ___ ___
1 2 3 4 5 6 7 8 9

#6 ___ ___ ___ ___ ___ ___ ___ ___ ___ ___
1 2 3 4 5 6 7 8 9 10

#7 ___ ___ ___ ___ ___ ___ ___ ___ ___ ___ ___
1 2 3 4 5 6 7 8 9 10 11

#8 ___ ___ ___ ___ ___ ___ ___ ___ ___ ___ ___
1 2 3 4 5 6 7 8 9 10 11

#9 ___ ___ ___ ___ ___ ___ ___ ___ ___ ___ ___
1 2 3 4 5 6 7 8 9 10 11

#10 ___ ___ ___ ___ ___ ___ ___ ___ ___ ___ ___
1 2 3 4 5 6 7 8 9 10 11

LETTERS MISSED

#1 |__|__|__|__|__|__|
#2 |__|__|__|__|__|__|
#3 |__|__|__|__|__|__|
#4 |__|__|__|__|__|__|
#5 |__|__|__|__|__|__|
#6 |__|__|__|__|__|__|
#7 |__|__|__|__|__|__|
#8 |__|__|__|__|__|__|
#9 |__|__|__|__|__|__|
#10 |__|__|__|__|__|__|

POSITION CHART

1 / *4*	**2** / *0*	**3** / *9*	**4** / *5, 7*	**5** / *0*
6 / *0*	**7** / *1, 3*	**8** / *0*	**9** / *0*	**10** / *5*
11 / *3, 6*	**12** / *0*	**13** / *4, 8*	**14** / *2, 9*	**15** / *0*
16 / *6*	**17** / *0*	**18** / *6, 9*	**19** / *4*	**20** / *5,9,11*
21 / *0*	**22** / *2, 5*	**23** / *0*	**24** / *0*	**25** / *5*
26 / *6, 10*	**27** / *0*	**28** / *8*	**29** / *0*	**30** / *3*
31 / *0*	**32** / *4, 10*	**33** / *0*	**34** / *0*	**35** / *0*
36 / *0*	**37** / *0*	**38** / *5, 7*	**39** / *0*	**40** / *7, 9*
41 / *0*	**42** / *1*	**43** / *0*	**44** / *5, 9*	**45** / *0*
46 / *11*	**47** / *0*	**48** / *3, 6*	**49** / *0*	**50** / *0*
51 / *2*	**52** / *0*	**53** / *0*	**54** / *0*	**55** / *2, 7*
56 / *0*	**57** / *1*	**58** / *0*	**59** / *2*	**60** / *0*
61 / *8*	**62** / *0*	**63** / *6, 8*	**64** / *0*	**65** / *0*
66 / *0*	**67** / *1, 4*	**68** / *0*	**69** / *0*	**70** / *8*
71 / *10*	**72** / *0*	**73** / *0*	**74** / *0*	**75** / *7*
76 / *0*	**77** / *0*	**78** / *3*	**79** / *0*	**80** / *0*
81 / *3*	**82** / *0*	**83** / *0*	**84** / *0*	**85** / *0*

LETTER CHART

	A	B	C	D	E	F	G	H	I	J	K	L	M	N	O	P	Q	R	S	T	U	V	W	X	Y	Z
#1	75	2	50	85	5	23	67	2	22	56	5	72	39	81	47	17	8	83	49	31	58	16	6	77	82	34
#2	68	33	72	8	24	2	83	36	53	41	84	11	50	37	59	69	50	41	61	5	38	67	24	64	5	53
#3	47	9	66	21	81	42	60	35	76	8	35	16	61	19	5	49	62	51	79	69	4	45	8	12	31	85
#4	23	58	76	67	10	82	66	49	12	2	79	23	70	16	30	6	41	34	84	27	55	27	64	9	17	45
#5	18	12	27	57	34	10	75	25	28	85	39	33	12	72	56	1	65	56	78	5	74	69	50	23	51	21
#6	65	53	33	76	14	79	61	52	10	73	29	52	74	26	75	7	8	62	1	43	80	36	82	43	80	39
#7	23	82	9	42	51	45	81	15	40	58	77	19	68	46	71	82	52	39	27	63	25	17	60	21	2	54
#8	75	34	24	84	59	35	15	66	32	64	84	61	9	27	80	42	29	78	20	16	73	31	77	37	85	74
#9	55	43	16	77	37	43	68	85	44	77	24	45	57	46	71	15	37	6	78	13	54	36	43	17	65	60
#10	21	67	29	47	36	83	70	52	35	56	15	54	3	41	22	73	31	48	46	33	71	6	83	29	75	62

17 Metabolism

☞ Chapter Outline/Objectives

Introduction

1. Define the terms metabolism and nutrition.

Metabolism of Absorbed Nutrients

2. Distinguish between anabolism and catabolism.
3. Name the molecule that represents stored chemical energy in the cell.
4. Use the terms glucose, anaerobic, cytoplasm, pyruvic acid, and ATP to describe glycolysis.
5. Use the terms pyruvic acid, acetyl CoA, citric acid cycle, mitochondria, and ATP to describe the aerobic phase of cellular respiration.
6. Define glycogen, glycogenesis, glycogenolysis, and gluconeogenesis.
7. List six uses of proteins in the body.
8. Explain how amino acids can be used for energy if there is insufficient carbohydrate or fat.
9. Describe the pathway by which fatty acids are broken down to produce ATP.
10. Name two key molecules in the metabolism and interconversion of carbohydrates, proteins, and fats.
11. Define the term Calorie.
12. List three uses of energy in the body.
13. Explain what is meant by basal metabolism and state four factors that influence it.
14. State the avenue of energy use that can be controlled voluntarily.
15. Define thermogenesis.

Basic Elements of Nutrition

16. List three functions of carbohydrates in the body.
17. State the number of Calories in one gram of pure carbohydrate.
18. Distinguish between simple sugars and complex carbohydrates.
19. Name three types of complex carbohydrates that are important in nutrition and give a source for each one.
20. Explain the importance of fiber in the diet.
21. State three categories of protein function in the body.
22. State the number of Calories in one gram of protein.
23. Distinguish between essential and nonessential amino acids and between complete and incomplete proteins.
24. Explain why two or more incomplete proteins should be eaten in the same meal.
25. List six functions of fats in the body.
26. State the number of Calories in one gram of fat.
27. Discuss the American Heart Association's recommendations for dietary intake of fats.
28. Identify vitamins as either water soluble or fat soluble.
29. Explain why vitamins are important in the diet.
30. List four uses of minerals in the body.
31. State five reasons water is necessary in the body.

Body Temperature

32. Distinguish between core temperature and shell temperature.
33. Explain three ways in which core temperature is maintained when the environmental temperature is cold.
34. State four ways in which heat is lost from the body.
35. Describe the general mechanism by which the body maintains core temperature and identify the region of the brain that integrates this mechanism.

☞ Learning Exercises

Introduction (Objective 1)

1. Write the terms that match the given definitions .

_______________________________ Total of all chemical reactions in the body

_______________________________ Chemical reactions within cells to produce energy

_______________________________ Substances that speed up physiologic reactions

_______________________________ Acquisition and utilization of nutrients

Metabolism of Absorbed Nutrients (Objectives 2-15)

1. Match the following terms with the correct definitions or descriptive phrases.

A. Acetyl CoA
B. Aerobic
C. Anaerobic
D. Beta oxidation
E. Cytoplasm
F. Deamination
G. Gluconeogenesis
H. Glucose
I. Glycogen
J. Glycogenesis
K. Glycogenolysis
L. Glycolysis
M. Lactic acid
N. Lipogenesis
O. Mitochondria

_______ Most important simple sugar in cellular metabolism

_______ Term that means oxygen is not required

_______ Reactions in which glucose is split into two molecules of pyruvic acid

_______ Fate of pyruvic acid in absence of oxygen

_______ Term that means oxygen is required

_______ Molecule that enters the citric acid cycle

_______ Storage form of glucose

_______ Location of glycolysis reactions

_______ Location of citric acid cycle reactions

_______ Conversion of glucose to glycogen

_______ Conversion of glucose to fat

_______ Conversion of glycogen into glucose

_______ Conversion of noncarbohydrates to glucose

_______ Principal reaction that prepares amino acids for use as an energy source

_______ Reactions that convert fatty acids to acetyl CoA

2. State six uses of the proteins that are synthesized in the body. Do not include their possible use as an energy source.

_______________ _______________

_______________ _______________

_______________ _______________

3. What are the two key molecules in the interconversion of carbohydrates, proteins, and fats?

_______________ _______________

4. Define or explain what is meant by the nutritional Calorie or kilocalorie.

____

____

5. List the three uses of energy in the body. Draw a circle around the one that accounts for most of the energy used. Place an asterisk (*) by the one that can be controlled voluntarily.

____

____

____

Basic Elements of Nutrition (Objectives 16-31)

1. Indicate whether each of the following most closely applies to carbohydrates, lipids, or proteins. Use C = carbohydrates, L = lipids, and P = proteins.

_______ Primary energy source

_______ Regulate body processes

_______ Major component of cell membranes

_______ Transport vitamins A, D, E, K

_______ Add bulk to the diet

_______ Fatty acids

_______ Provide insulation and protection

_______ Provide structure

_______ Glucose

_______ Concentrated energy

_______ Fiber

_______ Hormones, enzymes

_______ Steroids

_______ Glycogen

_______ Triglycerides

_______ Amino acids

2. How many Calories are in each of the following?

_______ 1 gram pure carbohydrate

_______ 1 gram pure protein

_______ 1 gram pure fat

_______ 5 grams carbohydrate

_______ 4 grams protein

_______ 3 grams fat

3. Complete the following paragraph by writing the correct words in the blanks.

   Amino acids that cannot be synthesized in the body and must be supplied in the diet are called ________________ amino acids . A protein that contains all of these amino acids is called a ________________ protein. Other proteins, called incomplete proteins, should be eaten in combinations to provide all of the ____________ amino acids.

4. Complete the following paragraph by writing the correct words in the blanks.

   The American Heart Association recommends that no more than __________of the daily calorie intake should be in the form of fats. Further, it recommends a reduction in the amount of saturated fats in the diet. In general, foods that are high in saturated fats are also high in ______. Cholesterol intake should be limited to less than __________ per day.

5. Classify the following vitamins as being either water soluble or fat soluble. Use W for water soluble and F for fat soluble.

   _______ Thiamine　　_______ Vitamin A　　_______ Vitamin D

   _______ Niacin　　_______ Vitamin C　　_______ Riboflavin

6. Indicate whether each of the following most closely applies to uses of vitamins, minerals, or water in the body. Use V = vitamins, M = minerals, and W = water.

   _______ Medium for chemical reactions　　_______ Incorporated into bones and teeth

   _______ Release energy from nutrients　　_______ Maintenance of body temperature

   _______ Nucleic acid synthesis　　_______ Regulation of body fluid levels

Body Temperature (Objectives 32-35)

1. Match the following terms with the correct definitions or descriptive phrases about body temperature.

   A. Blood
   B. Constriction of vessels in skin
   C. Core temperature
   D. Epinephrine
   E. Heat
   F. Hypothalamus
   G. Perspiration
   H. Radiation
   I. Shell temperature
   J. Shivering

   ____ Byproduct of cellular metabolism　　____ Two methods of increasing body heat

   ____ Location of the body's thermostat　　____

   ____ Temperature of internal organs　　____ Two methods of releasing body heat

   ____ Heat-producing hormone　　____

   ____ Distributes heat through the body　　____ Temperature at the body surface

☞ Chapter Self-Quiz

1. For each of the following, indicate whether it refers to anabolism or to catabolism. Use A = anabolism and C = catabolism.

    ___________ Releases energy

    ___________ Cellular respiration

    ___________ Dehydration synthesis

    ___________ Breaks large molecules into smaller ones

2. Which one of the following is not a characteristic of glycolysis? (a) anaerobic; (b) occurs in mitochondria; (c) produces ATP; (d) glucose is converted to pyruvic acid

3. Which one of the following is not a characteristic of the aerobic phase of cellular respiration? (a) occurs in mitochondria; (b) called the citric acid cycle; (c) end product is acetyl CoA; (d) releases energy in the form of ATP and heat

4. What large carbohydrate molecule represents the storage form of glucose in humans?

5. For each of the following, indicate whether it is a feature or characteristic of deamination or beta oxidation. Use D = deamination and B = beta oxidation.

    ___________ A reaction in protein catabolism

    ___________ Breaks off two-carbon segments from fatty acid chains

    ___________ Produces ammonia

    ___________ A reaction in the utilization of amino acids for energy

    ___________ Produces acetyl CoA

    ___________ A step in the conversion of amino acids to glucose

    ___________ A reaction in lipid catabolism

6. Which is greater: energy used for basal metabolism or energy used for thermogenesis?

7. How many Calories are derived from 8 grams of carbohydrate? How many grams of fat give the same number of Calories?

8. Match each description with the appropriate carbohydrate.

_____	A monosaccharide	A. fiber
_____	Nondigestible complex polysaccharide	B. glucose
_____	Formed from two glucose molecules	C. maltose
_____	Glucose storage form in plants	D. starch
_____	Table sugar	E. sucrose

9. Name the vitamin that is important for

_____ Collagen synthesis

_____ Formation of pigments in the retina

_____ Synthesis of clotting factors

_____ Formation of erythrocytes

_____ Synthesis of DNA

10. In response to a cold environment, (a) the cutaneous blood vessels dilate so you feel warmer; (b) the metabolic rate deceases to conserve heat; (c) core temperature decreases before the shell temperature decreases; (d) epinephrine and norepinephrine promote heat production

☞ Terminology Exercises

WORD PART	MEANING	WORD PART	MEANING
ana	up	neo-	new
-bol-	throw, put	nutri-	to nourish
cata-	down	pyr-	fever, fire
-gen-	producing	therm-	temperature
lys-	to take apart	-tion	process of
mal-	bad, poor	vita-	life

Use word parts given above or in previous chapters to form words that have the following definitions.

________________________________ Producing fever

________________________________ Taking apart glycogen

________________________________ Process of nourishing

________________________________ Producing new glucose

________________________________ Excessive temperature

Using the definitions of word parts given above or in previous chapters, define the following words.

Hydrolysis ____

Hypothermia ____

Lipogenesis ____

Antipyrogenic ____

Thermogenesis ____

Match each of the following definitions with the correct word.

____________ A substance required for life — A. catabolism

____________ Poor nourishment — B. deamination

____________ Throwing down or taking apart — C. lipolysis

____________ Taking apart fats — D. malnutrition

____________ Process of removing an amino group — E. vitamin

☞ Fun and Games

The object of this puzzle is to accumulate as many points as possible for the words you select as answers for the clues. To do the puzzle, answer each clue with a **single** word and write that word in the space by the clue. Each letter of the alphabet is assigned point values as indicated. Using these point values, add up your score for each answer. Each clue has more than one possible answer and you should try to choose the one that gives the highest point value. Finally, add the ten individual scores to get your total score for the puzzle. For fair play, use single word answers only and avoid answers, such as lactic acid and beta oxidation, that contain two words. Try competing with your classmates to see who can get the highest score! Have fun!!

A = 1	B = 2	C = 2	D = 2	E = 1	F = 3	G = 3
H = 3	I = 1	J = 5	K = 4	L = 2	M = 2	N = 1
O = 1	P =3	Q = 5	R = 1	S = 1	T = 1	U = 1
V = 4	W = 4	X = 5	Y = 3	Z = 5		

Clue	Single Word Answer	Points
1. One of the phases of metabolism	______________	____
2. Any process that breaks down or produces glucose	______________	____
3. A hexose monosaccharide	______________	____
4. A disaccharide	______________	____
5. A complex polysaccharide	______________	____
6. A lipid	______________	____
7. Mechanism of heat loss	______________	____
8. A mineral required in the diet	______________	____
9. One of the B vitamins	______________	____
10. Protein catabolism reaction	______________	____

18 Urinary System

☞ Chapter Outline/Objectives

Functions of the Urinary System

1. State six functions of the urinary system.

Components of the Urinary System

2. Describe the location of the kidneys.
3. Label the capsule, cortex, renal pyramids, renal columns, papillae, major and minor calyces, renal pelvis, and ureter on an illustration of a frontal section through the kidney.
4. Distinguish between the renal corpuscle and the renal tubule.
5. Draw a diagrammatic representation of a nephron and label the glomerulus, glomerular capsule, proximal convoluted tubule, loop of the nephron with descending and ascending limbs, and the distal convoluted tubule.
6. Identify the parts of a nephron that are in the cortex and those that are in the medulla.
7. Name the structures that collect filtrate from the nephrons and takes it to a minor calyx. Indicate where these structures are located in the kidney.
8. Name the two parts of the juxtaglomerular apparatus and state where they are located.
9. Trace the pathway of blood flow through the kidney from the renal artery to the renal vein.
10. Name the structure that transports urine away from the kidney.
11. Describe the location of the urinary bladder.
12. Give the name of the folds of the mucosa and the name of the smooth muscle in the wall of the urinary bladder.
13. Name the openings that form the three points of the trigone.
14. Define the term micturition.
15. Distinguish between the internal and external urethral sphincters with respect to location and type of muscle.
16. Compare the length of the female urethra with the length of the male urethra.

Urine Formation

17. List the three basic steps in urine formation.
18. State the direction of fluid flow in glomerular filtration.
19. Identify three different types of pressure that affect the rate of glomerular filtration and describe how these interact.
20. State the direction substances move in tubular reabsorption.
21. Distinguish between the mechanism for solute reabsorption and the mechanism for water reabsorption.
22. Explain why some substances, such as glucose, have limited reabsorption and what happens when concentration exceeds this limit.
23. Identify the portion of the nephron tubule that is not permeable to water.
24. Define tubular secretion and name four substances that are added to the urine during this process.
25. Explain how urine production has a role in maintaining blood concentration and volume.
26. Name two hormones that affect kidney function and explain the effect of each one.
27. Name the enzyme that stimulates the production of angiotensin II and is produced by the kidneys.
28. Describe two mechanisms by which angiotensin II increases blood pressure.

Characteristics of Urine

29. State the color, specific gravity, and pH of urine.
30. Describe the chemical composition of urine.
31. Name at least four abnormal constituents of urine that may indicate pathologic conditions.

Body Fluids

32. Compare the volume of intracellular fluid with the volume of extracellular fluid and the volume of interstitial fluid with the volume of intravascular fluid.
33. List three sources of fluid intake.
34. List four avenues of fluid loss from the body.
35. Identify three major ions in the extracellular fluid and two major ions in the intracellular fluid.
36. Identify the primary hormone that regulates electrolyte concentration.
37. State the normal pH range of the blood and the terms used to indicate deviations from the normal.
38. Describe three mechanisms for maintaining normal blood pH.

☞ Learning Exercises

Functions of the Urinary System (Objective 1)

1. What are six functions of the urinary system?

________________________ ________________________

________________________ ________________________

________________________ ________________________

Components of the Urinary System (Objectives 2-16)

1. Name the four components of the urinary system in the sequence urine flows through them.

____________ → ____________ → ____________ → ____________

2. Identify the parts of the kidney by matching the letters from the diagram with the correct structure in the list.

_______ Major calyx

_______ Minor calyx

_______ Renal capsule

_______ Renal column

_______ Renal cortex

_______ Renal papilla

_______ Renal pelvis

_______ Renal pyramid

_______ Ureter

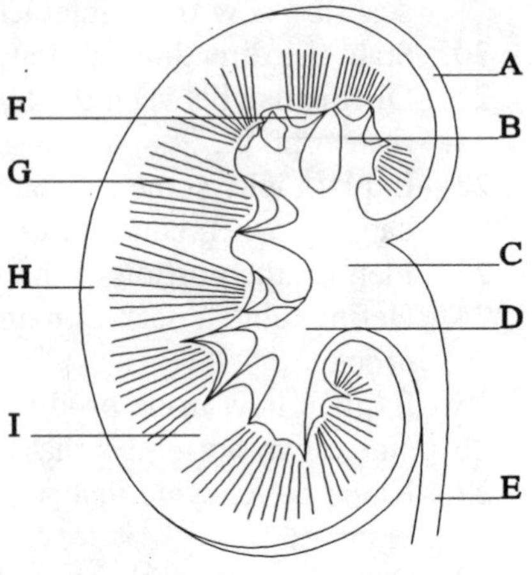

A
B
C
D
E
F
G
H
I

3. Identify the parts of a nephron by matching the letters from the diagram with the correct structure in the list. Draw a box around the renal corpuscle. Draw a circle around the region of the juxtglomerular apparatus. Use blue to color the portions that are located in the medulla of the kidney.

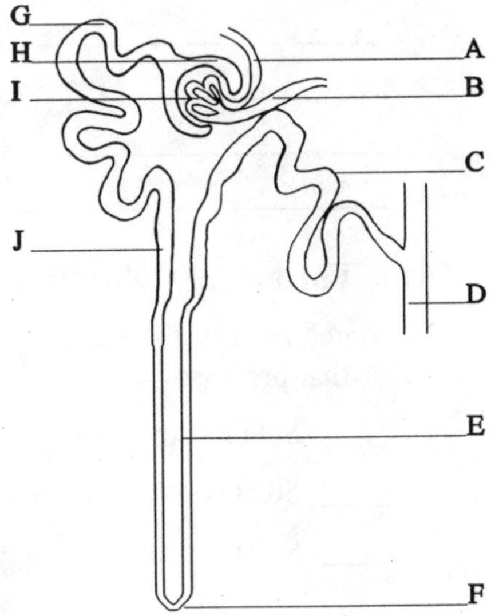

_______ Afferent arteriole
_______ Ascending limb
_______ Collecting duct
_______ Descending limb
_______ Distal convoluted tubule
_______ Efferent arteriole
_______ Glomerular capsule
_______ Glomerulus
_______ Nephron loop
_______ Proximal convoluted tubule

G H I A B C J D E F

4. Write the terms that match the following phrases.

_______________ Modified cells in the ascending limb
_______________ Modified cells in the afferent arteriole
_______________ Enzyme produced by juxtaglomerular cells
_______________ Structure that monitors NaCl in the urine
_______________ Rate of blood flow through the kidney
_______________ Arteries located in the renal sinus
_______________ Arteries located in the renal columns
_______________ Arteries that pass over the renal pyramids
_______________ Branches of the arcuate arteries
_______________ Vessel that carries blood to the glomerulus
_______________ Capillary network around renal tubules
_______________ Location of the interlobar vein
_______________ Veins that join to form the renal vein

5. Write the terms that match the following phrases about the ureters, urinary bladder, and urethra.

_______________ Transport urine from kidney to urinary bladder
_______________ Epithelium in mucosa of ureters
_______________ Temporary storage site for urine
_______________ Folds in mucosa of urinary bladder
_______________ Epithelium in mucosa of urinary bladder

_______________ Smooth muscle in wall of urinary bladder
_______________ Triangular region in floor of urinary bladder
_______________ Muscle type forming internal urethral sphincter
_______________ Muscle type forming external urethral sphincter
_______________ Transports urine from bladder to outside
_______________ Region of the male urethra nearest the bladder
_______________ Longest portion of the male urethra

Urine Formation (Objectives 17-28)

1. Match each of the three processes involved in urine formation with the correct description of that process.

____	Solutes pass from tubular cells into the filtrate	A. Glomerular filtration
____	Solutes pass from the glomerulus into the capsule	B. Tubular reabsorption
____	Solutes pass from tubules into peritubular capillaries	C. Tubular secretion

2. The following diagram illustrates a glomerulus with the glomerular capsule around it. Representative osmotic and hydrostatic pressures also are indicated. Draw red circles around the pressures that move substances from the glomerulus into the capsule and draw blue circles around the pressures that move substances from the capsule into the glomerulus. Calculate the net filtration pressure.

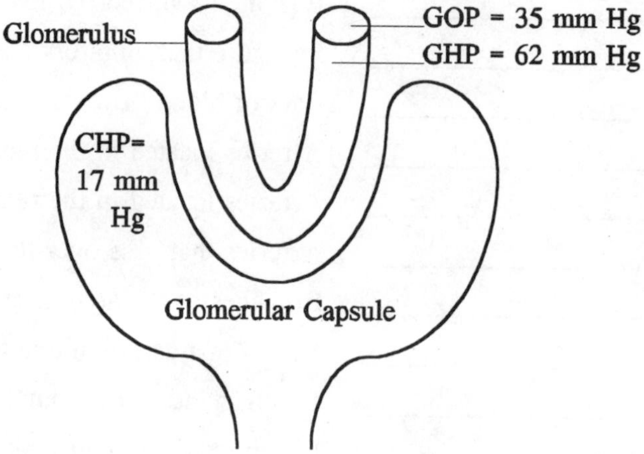

Glomerulus
GOP = 35 mm Hg
GHP = 62 mm Hg
CHP= 17 mm Hg
Glomerular Capsule

3. Write the terms that match the following phrases about regulation of urine concentration and volume.

_______________ Hormone that increases reabsorption of sodium
_______________ Hormone that increases reabsorption of water
_______________ Hormone that promotes sodium and water loss
_______________ Enzyme produced by juxtaglomerular cells
_______________ Powerful vasoconstrictor
_______________ Act of expelling urine from the urinary bladder

4. Complete the following paragraph about tubular reabsorption and secretion by filling in the correct terms.

1) ____________________

2) ____________________

3) ____________________

4) ____________________

5) ____________________

6) ____________________

7) ____________________

8) ____________________

Tubular reabsorption moves substances from the filtrate into the blood and __1__ the volume of urine. Water is reabsorbed by the process of __2__. Most of the solutes are reabsorbed by __3__ mechanisms. Reabsorption of some solutes is limited by __4__. When the concentration of these solutes exceeds __5__ the excess appears in the __6__. Tubular __7__ adds substances to the urine. To help regulate blood pH, __8__ ions are added to the urine and removed from the body by this process.

5. In what two ways does angiotensin II act on the kidneys?

   A.

   B.

Characteristics of Urine (Objectives 29-31)

1. Place a check (✓) before each of the following statements that is correct.

_______ The color of urine is due to urochrome.

_______ Urine is usually slightly acidic, but may be alkaline.

_______ Normal urine volume is generally about three liters per 24 hours.

_______ High protein diets tend to make urine more alkaline.

_______ The specific gravity of urine is slightly greater than 1.

_______ Solutes make up about 20 % of the urine volume.

_______ The predominant solute in urine is urea.

_______ In addition to urea, solutes in urine include glucose and albumin.

2. Write the clinical term that is used to indicate the presence of the following substances in the urine.

_______________ White blood cells   _______________ Albumin

_______________ Glucose   _______________ Erythrocytes

Body Fluids (Objectives 32-38)

1. List three types of sources for fluid intake. Place an asterisk (*) by the one that normally accounts for the greatest amount of intake.

   A.

   B.

   C.

2. List four avenues of fluid loss and place an asterisk (*) by the one that normally acounts for the greatest amount of loss.

   A. C.

   B. D.

3. The following boxes represent the volume of different fluid compartments in the body. Match the letter of each box with the correct name. Also indicate the percent of total body weight that is contained in each compartment.

Compartment	Letter	Percent
Extracellular	______	______
Interstitial	______	______
Intracellular	______	______
Plasma	______	______
Solutes	______	______
Total fluids	______	______

A

B

C

D

E

F

4. Write the terms that match the following phrases about electrolytes in the body.

   ______________________ Predominant cation in extracellular fluid

   ______________________ Predominant anion in extracellular fluid

   ______________________ Predominant cation in intracellular fluid

   ______________________ Predominant anion in intracellular fluid

   ______________________ Primary hormone that regulates electrolytes

5. Write the terms that match the following phrases about acid-base balance in the body.

   ______________________ Normal pH range of the blood

   ______________________ Clinical term for higher than normal pH

   ______________________ Clinical term for lower than normal pH

   ______________________ Three primary regulators of blood pH

   ____

   ____

☞ Chapter Self-Quiz

1. The functional unit of the kidney is the (a) pyramid; (b) collecting duct; (c) renal corpuscle; (d) nephron

2. The portion of a nephron that is located in the medulla is the (a) glomerulus; (b) nephron loop of Henle; (c) proximal convoluted tubule; (d) glomerular capsule; (e) distal convoluted tubule

3. Renin is produced by specialized cells in the (a) medulla; (b) renal pelvis; (c) afferent arteriole; (d) macula densa; (e) glomerulus

4. The arteries that are located in the renal columns are the (a) interlobar arteries; (b) arcuate arteries; (c) segmental arteries; (d) interlobular arteries; (e) renal arteries

5. Name the following:

______________________ Smooth muscle in the wall of the urinary bladder

______________________ Triangular region in the floor of the urinary bladder

______________________ Voluntary sphincter of the urethra

______________________ Proximal or first portion of the male urethra

6. The following represents a nephron tubule and a blood capillary. Draw arrows to indicate whether substances move from the capillary into the tubule or from the tubule into the capillary in each of the three steps in urine formation.

	Glomerular Filtration	Tubular Reabsorption	Tubular Secretion
Capillary			
Tubule			

7. Indicate whether each of the following factors increases or decreases the volume of urine.

_______________ Decreased capillary osmotic pressure

_______________ Increased tubular reabsorption of water

_______________ Decreased secretion of ADH

_______________ Increased secretion of aldosterone

_______________ Increased secretion of renin

8. Assume an adult male weighs 80 kg.

How many liters of water are in his body?

How many liters of intracellular fluid are there?

How many liters of interstitial fluid are there?

9. Which of the following is the primary extracellular cation? (a) sodium; (b) potassium; (c) chloride; (d) phosphate; (e) bicarbonate

10. Which of the following blood pH values represents acidosis? (a) 7.5; (b) 7.4; (c) 7.3; (d) none of the above; (e) all of the above

☞ Terminology Exercises

WORD PART	MEANING	WORD PART	MEANING
-atresia	without an opening	mict-	to pass
caly-	small cup	neph-	kidney
-chrom-	color, pigment	noct-	night
-continence	to hold	peri-	around
cyst-	bladder	-pexy	fixation
-ectasy	dilation	-phraxis	to obstruct
		pyel-	renal pelvis
-etic	pertaining to	ren-	kidney
juxta-	near to	-rrhaphy	suture
lith-	stone	-ur-	urine

Use word parts given above or in previous chapters to form words that have the following definitions.

_______________________ Urination at night

_______________________ Surgical fixation of the bladder

_______________________ Kidney stone

_______________________ Process of passing urine

_______________________ Near the glomerulus

Using the definitions of word parts given above or in previous chapters, define the following words.

Cystorrhaphy _______________________

Hematuria _______________________

Incontinence _______________________

Urethratresia _______________________

Urethrophraxis _______________________

Match each of the following definitions with the correct word.

_________ Inflammation around the ureter — A. cystectasy

_________ Glandular tumor of the kidney — B. dysuria

_________ Surgical removal of stone from renal pelvis — C. nephroadenoma

_________ Painful urination — D. periureteritis

_________ Dilation of the urinary bladder — E. pyelolithotomy

☞ Fun and Games

Each of the puzzles below consists of a series of clues for which the answers "add-up" to solve the puzzle. The individual clues are followed by a series of blanks with numbers. When you have determined the answer to a clue, transfer the letters to the correspondingly numbered spaces for the "total" or final answer. If there is no number under a blank, that letter does not appear in the final answer.

CLUES	ANSWERS
Vessel leading into the glomerulus	16 11 11 5 10 5 20 17 9 6 14 5 10 12 19 8 5
Reduced urine output	3 13 18 1 7 15 12 9
Normal fluid intake is about 2500 ______	4 2

TOTAL: First process in urine formation

1 2 3 4 5 6 7 8 9 10   11 12 13 14 15 16 17 18 19 20

CLUES	ANSWERS
Convoluted tubule preceding nephron loop	17 11 8 3 _ 9 14 7
Fluid portion of blood	18 13 5 24 9 16
Prefix meaning "near to"	1 23 3 22 16
Root meaning "without an opening"	19 4 15 10 24 _ 21
Folds in bladder wall	20 2 6 21 10
Twenty-first letter of the alphabet	12

TOTAL: Structure that monitors blood flow and secretes renin

1 2 3 4 5 6 7 8 9 10 11 12 13 14 15   16 17 18 19 20 21 22 23 24

CLUES	ANSWERS
Outer region of kidney	6 _ 13 3 7 2
Vessel leading away from glomerulus	1 14 14 7 4 1 _ 3 5 13 3 1 4 17 _ 8 1
Inner region of kidney	_ 7 18 10 9 15 12
Indentation on medial side of kidney	_ 17 11 16 _

TOTAL: Makes up 20% of the adult body weight

1 2 3 4 5 6 7 8 9 10 11 12 13   14 15 16 17 18

19 Reproductive System

☞ Chapter Objectives

Male Reproductive System

1. Distinguish between primary and secondary reproductive organs.
2. Label the parts of the male reproductive system on a diagram of a midsagittal section through the pelvis.
3. Describe the structure and location of the testes.
4. State the significance of seminiferous tubules and interstitial cells.
5. Draw and label a diagram or flow chart that illustrates spermatogenesis.
6. Name the process by which spermatids become spermatozoa and describe the three regions of a mature sperm.
7. Trace the pathway of sperm from the testes to the outside of the body.
8. Name three accessory glands of the male reproductive system and describe the contribution each makes to the seminal fluid.
9. Distinguish between the emission and ejaculation of semen.
10. Use the terms corpora cavernosa, corpus spongiosum, root, body, and glans penis to describe the structure of the penis.
11. Outline the physiological events in the male sexual response.
12. Describe the role of GnRH, FSH, LH, and testosterone in the male reproductive system.

Female Reproductive System

13. Label the parts of the female reproductive system on a diagram of a midsagittal section through the pelvis.
14. Describe the location and structure of the ovaries.
15. Draw and label a diagram or flow chart that illustrates oogenesis.
16. Compare oogenesis and spermatogenesis on the basis of when the different stages occur and the final results.
17. Describe the development of ovarian follicles as they progress from primordial follicles to primary follicles, secondary follicles, vesicular follicles, corpus luteum, and finally, the corpus albicans.
18. Identify the stage of oogenesis that exists at the time of ovulation.
19. Use the terms infundibulum and fimbriae in a description of the uterine tubes.
20. Use the terms fundus, body, cervix, internal os, external os, broad ligament, and anteflexed to describe features of the uterus.
21. Name the three layers of the uterine wall and describe the tissue in each one; distinguish between the two parts of the endometrium.
22. State three functions of the vagina.
23. Name six structures included in the vulva.
24. Name two accessory glands of the female reproductive system that are associated with the vestibule.
25. Outline the physiological events in the female sexual response.
26. Describe the roles of GnRH, FSH, LH, estrogen, and progesterone in the female reproductive system.
27. Define the terms menarche and menopause.
28. Name the three phases of the ovarian cycle, state when each phase occurs, and describe what happens in each phase.

29. Name the three phases of the uterine cycle, state when each phase occurs, and describe what happens in each phase.
30. Explain how the events of the ovarian cycle affect the uterine cycle.
31. Describe the hormonal changes that occur during menopause.
32. Describe the location and structure of the mammary glands.
33. Describe the effects of estrogen, progesterone, prolactin, and oxytocin on the mammary glands.

☞ Learning Exercises

Male Reproductive System (Objectives 1-12)

1. Identify the parts of the male reproductive system by matching the letters from the diagram with the correct structure in the list.

_______ Bulbourethral gland
_______ Corpus cavernosum
_______ Corpus spongiosum
_______ Ductus deferens
_______ Ejaculatory duct
_______ Epididymis
_______ Glans penis
_______ Prostate
_______ Scrotum
_______ Symphysis pubis
_______ Testicle
_______ Urethra (2 places)
____
_______ Urinary bladder

2. Number the following structures in the correct sequence, from 1 to 10, to trace the pathway of sperm from the seminiferous tubule to the exterior. Start with #1 for the seminiferous tubule.

_______ Ductus deferens
_______ Efferent ducts
_______ Ejaculatory duct
_______ Epididymis
_______ Membranous urethra
_______ Penile urethra
_______ Prostatic urethra
_______ Rete testis
_______ Seminiferous tubules
_______ Straight tubules

3. Write the terms that match the following phrases about the male reproductive system.

______________________	Male gonad
______________________	Male gamete
______________________	Pouch of skin that contains the male gonad
______________________	Smooth muscle in subcutaneous tissue of scrotum
______________________	Skeletal muscle fibers in the scrotum
______________________	Fibrous connective tissue capsule of the testes
______________________	Specific location of spermatogenesis
______________________	Cells that produce male sex hormones
______________________	Maturation of spermatids into spermatozoa
______________________	Region of spermatozoa that contains mitochondria
______________________	Number of chromosomes in a spermatid
______________________	Cells that nourish developing spermatids

4. Match each of the following phrases with the accessory gland to which it pertains.

A. Seminal vesicles B. Prostate gland C. Bulbourethral gland

______	Product secreted into the penile urethra
______	Encircles the proximal portion of the urethra
______	Located posterior to the urinary bladder
______	Secretion accounts for 60 % of the seminal fluid
______	Located near the base of the penis
______	Smallest of the accessory glands
______	Secretion has a high fructose content

5. Complete the following paragraph about the male sexual response by filling in the correct terms.

1) ________________
2) ________________
3) ________________
4) ________________
5) ________________
6) ________________
7) ________________
8) ________________
9) ________________
10) ________________
11) ________________

During sexual stimulation __1__ impulses dilate the __2__ and constrict the __3__ of the penis. This causes the spaces in the erectile tissue to fill with blood and the penis enlarges and becomes rigid. This is called __4__. With continued stimulation, these reflexes become more intense until they prompt a surge of __5__ impulses to the genital organs. These impulses cause the forceful discharge of semen into the urethra. This is called __6__. __7__, the forceful expulsion of semen to the exterior, immediately follows. The muscular contractions that result in the expulsion of semen are accompanied by feelings of pleasure, __8__ heart rate, __9__ blood pressure, and __10__. Collectively, these physiological responses are referred to as __11__.

6. Write the terms that match the following phrases about the hormones that control male reproductive functions.

_______________ Hypothalamic hormone that initiates puberty
_______________ Hormone from the interstitial cells
_______________ Hormone that stimulates the interstitial cells
_______________ Collective term for male sex hormones
_______________ With testosterone it stimulates spermatogenesis
_______________ Anterior pituitary hormones that act on testes (2)
____
_______________ Source of testosterone before birth
_______________ Two functions of testosterone
____

Female Reproductive System (Objectives 13-33)

1. Identify the parts of the female reproductive system by matching the letters from the diagram with the correct structure in the list.

_____ Cervix
_____ Clitoris
_____ Infundibulum
_____ Mons pubis
_____ Ovary
_____ Rectum
_____ Symphysis pubis
_____ Urethra
_____ Urinary bladder
_____ Uterine tube
_____ Uterus
_____ Vagina

2. Write the terms that match the following phrases about the internal female reproductive system.

_______________ Female gonad
_______________ Stage of oogenesis present at birth
_______________ Stage of oogenesis that is ovulated each month
_______________ Clear glycoprotein membrane around an oocyte

_______________ Cells around oocyte at ovulation

_______________ Develops from follicle after ovulation

_______________ Fingerlike extensions of the uterine tubes

_______________ Opening from the cervix into the vagina

_______________ Largest ligament that stabilizes the uterus

_______________ Muscular layer of the uterus

_______________ Endometrial region shed during menstruation

_______________ Mucous membrane covering vaginal orifice

3. Identify structures associated with the ovary and uterine tube by matching the letters from the diagram with the correct item in the list.

_______ Antrum

_______ Corpus albicans

_______ Corpus luteum

_______ Fimbriae

_______ Infundibulum

_______ Primary follicle

_______ Secondary follicle

_______ Secondary oocyte

_______ Uterine tube

_______ Vesicular follicle

4. Write the terms that match the following phrases about the female external genitalia.

_______________ Collective term for female external genitalia

_______________ Fat-filled folds of skin that enclose the genitalia

_______________ Mound of fat that overlies the symphysis pubis

_______________ Area between the two labia minora

_______________ Female structure homologous to male penis

_______________ Glands adjacent to the urethral orifice

_______________ Glands adjacent to the vaginal orifice

5. Write T before the true statements and F before the false statements about the female sexual response.

_______ The female sexual response consists of erection and orgasm.

_______ Sympathetic responses to sexual stimuli produce increased blood flow to erectile tissue.

_______ Sympathetic responses produce rhythmic contractions of the uterus and pelvic floor, accompanied by feelings of intense pleasure.

6. Match the following statements and phrases with the correct hormone. Some may have more than one correct response. Give all correct answers.

A. Gonadotropin-releasing hormone
B. Follicle-stimulating hormone
C. Luteinizing hormone
D. Estrogen
E. Progesterone

_______ Starts the events of puberty
_______ Stimulates growth of ovarian follicles
_______ Secreted by the hypothalamus
_______ Secreted by the anterior pituitary
_______ Secreted by cells of the ovarian follicle
_______ Secreted by the corpus luteum
_______ Triggers ovulation
_______ Stimulates secretory phase of uterine cycle
_______ Stimulates proliferative phase of uterine cycle
_______ Stimulates development of the corpus luteum
_______ Stimulates development of glandular tissue in the breast
_______ Stimulates development of the duct system in the breast
_______ Causes an accumulation of adipose tissue in the breast
_______ Levels increase after menopause
_______ Levels decrease after menopause

7. Complete the following statements by using the word **increases (I)** or **decreases (D)**.

_______ As ovarian follicles grow in response to FSH, estrogen secretion _______.
_______ Blood estrogen level _______ during days 6-13 of the menstrual cycle.
_______ After ovulation progesterone secretion _______.
_______ An increasing level of progesterone _______ LH secretion.
_______ The menstrual phase of the uterine cycle is caused by _______ in progesterone levels.

8. Write the terms that match the following phrases about the breast.

_______________ Circular pigmented area around the nipple
_______________ Bands of connective tissue that support the breast
_______________ Collects the milk from the glandular units
_______________ Dilated region of duct that is a reservoir for milk
_______________ Hormone that stimulates milk production
_______________ Hormone that causes ejection of milk from glands

☞ Chapter Self-Quiz

1. Which of the following best describes a difference between spermatogenesis and oogenesis? (a) a primary spermatocyte produces four spermatids but a primary oocyte produces only two ova; (b) primary spermatocytes all develop after puberty but primary oocytes all develop before birth; (c) secondary spermatocytes have 46 chromosomes but secondary oocytes have only 23; (d) LH stimulates spermatogenesis but FSH stimulates oogenesis

2. Arrange the given ducts in the correct sequence for the passage of sperm: (1) ejaculatory duct; (2) epididymis; (3) urethra; (4) ductus deferens; (5) efferent duct. (a) 2, 4, 1, 3, 5; (b) 4, 5, 2, 1, 3; (c) 5, 2, 4, 3, 1; (d) 5, 2, 4, 1, 3

3. Indicate whether each of the following phrases best describes the seminal vesicles (S), the prostate (P), or the bulbourethral glands (B).

   _____ Encircles the urethra

   _____ Contributes about 60% of the seminal fluid volume

   _____ Empties into the penile urethra

   _____ Secretion is thin and milky colored

   _____ Empties into the ejaculatory duct

4. Place an X before the following phrases that describe the corpus spongiosum.

   _____ Erectile tissue

   _____ Encircles the urethra

   _____ Two columns

   _____ Dorsal

   _____ Makes up the glans penis

5. Match the following hormones with their actions. There is only one best response for each action.

   A. FSH
   B. GnRH
   C. LH
   D. estrogen
   E. progesterone
   F. testosterone

   _____ Stimulates the interstitial cells

   _____ Hormone from the hypothalamus

   _____ Secreted by the ovary during the follicular phase

   _____ Stimulates production of testosterone

   _____ Promotes glandular secretion in the endometrium

_____ Stimulates secretion of FSH

_____ Promotes repair of uterine lining after menstruation

_____ Stimulates the seminiferous tubules

_____ Promotes male secondary sex characteristics

_____ With testosterone, stimulates spermatogenesis

6. What noncellular layer surrounds the oocyte at the time of ovulation? (a) corona radiata; (b) zona pellucida; (c) granulosa; (d) antrum

7. The portion of the uterine wall that changes during the uterine cycle is the (a) myometrium; (b) perimetrium; (c) stratum basale; (d) stratum functionale

8. At what time in the ovarian cycle is the secretion of LH at a maximum?

9. Which one of the following is not true about the vulva? (a) labia majora have an abundance of adipose; (b) labia minora form the prepuce over the clitoris; (c) the clitoris is the most anterior structure in the vestibule; (d) the opening for the vagina is anterior to the urethra

10. What effect does each of the following hormones have on the mammary glands?

    a. Estrogen

    b. Progesterone

    c. Prolactin

    d. Oxytocin

☞ Terminology Exercises

WORD PART	MEANING	WORD PART	MEANING
andr-	male	mamm-	breast
balan-	glans	meno-	month
colpo-	vagina	metr-	uterus
crypt-	hidden	oo-	egg, ovum
ejacul-	to shoot forth	oophor-	ovary
fimb-	fringe	orch-	testicle
follic-	small bag	prostat-	prostate
gynec-	female	-rrhagia	to burst forth
hyster-	womb or uterus	salping-	tube
labi-	lip	spadias	an opening

Use word parts given above or in previous chapters to form words that have the following definitions.

______________________ Study of the female

______________________ Inflammation of the glans penis

______________________ Producing maleness

______________________ Muscular layer of the uterus

______________________ Surgical repair of the breast

Using the definitions of word parts given above or in previous chapters, define the following words.

Hysterectomy ______________________

Salpingitis ______________________

Prostatomegaly ______________________

Oophorectomy ______________________

Anorchism ______________________

Match each of the following definitions with the correct word.

________	Urethra opens on dorsum of penis	A. colpoperineoplasty
________	Surgical repair of the vagina and perineum	B. colporrhaphy
________	Excessive monthly flow	C. dysmenorrhea
________	Suture of the vagina	D. epispadias
________	Difficult or painful monthly flow	E. menorrhagia

☞ Fun and Games

Some of the following statements are true and others are false. If the statement is true, circle the letter in the **true** column. If the statement is false, circle the letter in the **false** column. When you have finished, unscramble the circled letters in the true column to form a word that indicates the subject of this chapter. Do the same for the letters in the false column to find a second related word .

True	False	Statements
N	A	Sperm are formed within the seminiferous tubules of the testes.
E	N	Sustentacular cells secrete testosterone.
O	U	The mitochondria of the sperm are located in the tail piece where they provide energy for the flagellum.
R	C	Sperm production is a continuous process that begins at puberty and continues throughout the life of a male.
D	T	The urethra is longer in males than in females.
C	I	Untreated cryptorchidism results in male sterility.
S	O	Fluid from the prostate contains fructose and makes up about 60% of the semen.
M	I	The male urethra is surrounded by corpus cavernosum in the penis.
O	B	Erection of the penis results when the venous sinuses in the erectile tissue become engorged with blood.
W	Y	The forceful discharge of semen into the urethra is called ejaculation.
E	D	FSH stimulates seminiferous tubules in males and ovarian follicles in females.
G	T	Four secondary oocytes develop from each oogonium during meiosis.
O	J	The corpus luteum secretes estrogen in addition to progesterone.
K	N	The oviduct is continuous with the ovary at one end and the uterus at the other.
T	L	The uterus normally is anteflexed over the superior surface of the urinary bladder.
P	S	The stratum basale of the endometrium functions to rebuild the stratum functionale after menstruation.
I	U	The thickest portion of the uterine wall is the myometrium.
A	T	The urethra and vagina open into a region called the vulva.
U	B	Parasympathetic responses to sexual stimuli produce increased blood flow to the clitoris.
D	C	The follicular phase of the ovarian cycle corresponds to the secretory phase of the uterine cycle.
G	I	Spermatozoa and ova are approximately the same size.
R	L	The hypothalamus secretes releasing factors that regulate, to some extent, the secretion of gonadotropic hormones from the pituitary gland.

True Letters______________________________ **Word**______________________________

False Letters______________________________ **Word**______________________________

20 Development

☞ Chapter Outline/Objectives

Fertilization

1. Define prenatal development and and postnatal development.
2. Define the term capacitation.
3. Describe the events in the process of fertilization, state where it normally occurs, and name of the cell that is formed as a result of fertilization.
4. Name the three divisions of prenatal development and state the period of time for each.

Preembryonic Period (Two weeks)

5. Describe three significant developments that take place during the preembryonic period.
6. Define the terms cleavage, blastomeres, morula, blastocyst, blastocele, trophoblast, and inner cell mass.
7. Describe where implantation normally occurs and how it saves itself from being aborted.
8. Name the three primary germ layers.

Embryonic Development (Six weeks)

9. Describe three significant developments that take place during the embryonic period.
10. Name the four extraembryonic membranes and describe the function of each one.
11. Describe the formation, structure, and functions of the placenta.
12. List five derivatives from each of the primary germ layers.

Fetal Development (Thirty weeks)

13. State the two fundamental processes that take place during fetal development.
14. Name, describe the location, and state the function of five structures that are unique in the circulatory pattern of the fetus.

Parturition and Lactation

15. Define the terms gestation, parturition, and labor.
16. State the length of the normal gestation period from the time of the last menstrual period and from the time of fertilization.
17. Describe the roles of the hypothalamus, estrogen, progesterone, oxytocin, and prostaglandins in promoting labor.
18. Describe the three stages of labor.
19. Describe the changes that take place in the baby's respiratory system and circulatory pathway at birth or soon after birth.
20. Distinguish between colostrum and milk.
21. Explain the relationship between a baby's suckling, the mother's hypothalamus, and milk production and ejection.

Postnatal Development

22. Name and define six periods in postnatal development.

☞ Learning Exercises

Fertilization (Objectives 1-4)

1. Write the terms that match the following phrases about fertilization.

_______________ Single cell that is product of fertilization

_______________ Cells that surround ovulated secondary oocyte

_______________ Process that weakens acrosomal membrane

_______________ Length of time ovulated oocytes are fertile

_______________ Usual site of fertilization

Preembryonic Period (Objectives 5-8)

1. Write the terms that match the following phrases about the preembryonic period.

_______________ Three developments during preembryonic period

____

____

_______________ Early cell divisions of the zygote

_______________ Cells that are the result of cleavage

_______________ Solid ball of cells resulting from cell division

_______________ Hollow sphere of cells formed by fifth day

_______________ Cluster of cells that becomes embryo

_______________ Cavity within hollow sphere of cells

_______________ Flattened cells around cavity of blastocyst

_______________ Blastocyst cells that contribute to placenta

_______________ Layer of uterine wall where implantation occurs

_______________ Hormone secreted by the blastocyst

_______________ Hormone that maintains the uterine lining

_______________ Cluster of cells that forms primary germ layers

_______________ Three primary germ layers

____

____

Embryonic Development (Objectives 9-12)

1. Complete the following statements by writing the correct words in the blanks.

   The period of embryonic development lasts from the beginning of the ____________ week after conception to the end of the ________________ week. Three significant developments during this period are the formation of the ______________________ membranes, formation of the __________________, and formation of all the body ______________________ systems. During this period the developing offspring is called an ______________________.

2. Match each of the following descriptive phrases with the correct extraembryonic membrane from the list.

   A. Amnion
   B. Allantois
   C. Chorion
   D. Yolk sac

   _______ Forms a sac around the developing embryo
   _______ Produces the primordial germ cells
   _______ Becomes part of the umbilical cord
   _______ Develops from the trophoblast
   _______ Cushions and protects developing offspring
   _______ Contributes to the formation of the placenta
   _______ Develops fingerlike projections called villi
   _______ Filled with fluid

3. Complete the following paragraph about the placenta by filling in the correct terms.

   1) ____________________
   2) ____________________
   3) ____________________
   4) ____________________
   5) ____________________
   6) ____________________
   7) ____________________
   8) ____________________
   9) ____________________

   The placenta develops as __1__ from the embryo penetrate the __2__ of the uterus. The __3__ become highly vascular and extend to the __4__ arteries and veins. The spaces in the endometrium are filled with maternal __5__. This interface allows __6__ and nutrients to diffuse from the mother's blood into the fetal blood. Metabolic wastes and __7__ diffuse from the __8__ into the __9__.

4. Match each of the following tissues with the primary germ layer from which it is derived.

A. Ectoderm B. Mesoderm C. Endoderm

_______ Epithelial lining of the digestive tract

_______ Epidermis of the skin

_______ Cardiac muscle

_______ Nervous tissue

_______ Cartilage

_______ Hair, nails, glands of the skin

_______ Respiratory epithelium

_______ Epithelium lining the blood vessels

_______ Lining of the oral cavity

_______ Bone

_______ Dermis of the skin

_______ Epithelial lining of the vagina

Fetal Development (Objectives 13 and 14)

1. Write the terms that match the following phrases about fetal development.

_______________________ First recognizable movements of fetus

_______________________ Protective coating over fetal skin

_______________________ Fine hair that covers fetal body

_______________________ Opening between the atria in the fetus

_______________________ Transports blood from placenta to fetus

_______________________ Allows blood to bypass fetal liver

Parturition and Lactation (Objectives 15-21)

1. Write the terms that match the following phrases about labor and devlivery.

_______________ Process of giving birth to an infant

_______________ Series of contractions to expel fetus

_______________ Hormone that inhibits uterine contractions

_______________ Sensitizes uterus to effects of oxytocin

_______________ Secretes oxytocin

_______________ Act with oxytocin to stimulate contractions

_______________ Type of feedback between oxytocin and uterus

_______________ Longest stage of labor

_______________ Stage of labor in which fetus is delivered

_______________ Stage characterized by rhythmic contractions

_______________ Final stage of labor

_______________ Normal position, or presentation, of baby

2. Complete the following paragraph about the changes that take place in the lungs immediately after birth.

1) _______________

2) _______________

3) _______________

4) _______________

5) _______________

6) _______________

The fetal lungs are __1__ and nonfunctional. When the __2__ is cut, the oxygen supply from the mother ceases. Increasing __3__ levels, decreasing __4__, and __5__ oxygen stimulate the respiratory center in the __6__. The respiratory muscles contract and the baby takes its first breath. Usually this is strong and deep and inflates the alveoli.

3. Write the terms that match the following phrases about lactation.

_______________ Refers to production and ejection of milk

_______________ Hormone that stimulates milk production

_______________ Hormone that stimulates milk ejection

_______________ Inhibit milk production during pregnancy (2)

____

_______________ Yellowish fluid secreted before milk begins

4. Complete the following paragraph about the stimulation for milk production.

1) _______________
2) _______________
3) _______________
4) _______________
5) _______________
6) _______________

Each time a mother nurses her infant, impulses from the nipple to the __1__ stimulate the release of __2__. This causes a temporary increase in __3__, which stimulates __4__ for the next nursing period. If a mother stops nursing her baby, milk production __5__ within a few __6__.

Postnatal Development (Objective 22)

1. Identify the period of postnatal development that is described by each of the following .

_______________ Lasts for about a month after birth

_______________ Period of old age

_______________ Lasts from the end of first year until puberty

_______________ Lasts from end of first month to end of first year

_______________ Puberty until adulthood

_______________ Body weight generally triples

_______________ Bladder and bowel controls are established

_______________ Degenerative changes become significant

☞ Chapter Self-Quiz

1. Define the following terms:

   Capacitation

   Zygote

   Cleavage

   Chorionic villi

   Senescence

2. Arrange the following events in the correct sequence: (1) formation of the inner cell mass; (2) cleavage; (3) appearance of lanugo hair; (4) formation of the primary germ layers; (5) heart starts beating. (a) 1, 2, 4, 3, 5; (b) 1, 2, 5, 3, 4; (c) 4, 1, 2, 5, 3; (d) 2, 1, 4, 5, 3; (e) 2, 1, 4, 3, 5

3. Name the primary germ layer from which each of the following is derived.

   ______________________ Lens of the eye

   ______________________ Dermis of the skin

   ______________________ Lining of the digestive tract

   ______________________ Cartilage and bone

   ______________________ Epidermis of the skin

4. Name the extraembryonic membrane that does each of the following,

   ______________________ Functions in the formation of the placenta

   ______________________ Contributes to the development of umbilical vessels

   ______________________ Source of primordial germ cells

   ______________________ Provides a sac of fluid that allows freedom of movement

   ______________________ Produces blood for the embryo

5. The vessel between the umbilical vein and the inferior vena cava is the (a) umbilical artery; (b) ductus venosus; (c) ductus arteriosus; (d) internal iliac artery

6. If a woman is 45 days past the beginning of her last menstrual period, the developing offspring is (a) in the cleavage stage of development; (b) in the preembryonic stage of development; (c) in the embryonic stage of development; (d) in the fetal stage of development; (e) in the implantation stage of development

7. Which of the following is <u>not</u> true about the placenta? (a) it forms from both embryonic and maternal tissue; (b) blood-filled lacunae in the endometrium surround chorionic villi; (c) maternal blood enters the chorionic villi to provide oxygen for the fetus; (d) it is expelled in the placental stage of labor

8. The birth of an infant is called (a) gestation; (b) parturition; (c) labor; (d) lactation.

9. Which of the following hormones is most responsible for continued milk production when breast feeding an infant? (a) human chorionic gonadotropin; (b) placental lactogen; (c) estrogen; (d) prolactin; (e) oxytocin

10. Which of the following does <u>not</u> refer to the neonatal period? (a) begins at the moment of birth; (b) temperature-regulating mechanisms may not be fully developed; (c) baby learns to smile and laugh; (d) foramen ovale normally closes

☞ Terminology Exercises

WORD PART	MEANING
amnio-	a fetal membrane, a lamb
-cente	to puncture surgically
cleav-	to divide
cyesi-	pregnancy
cyst-	bag, hollow
galacto-	referring to milk
gravid-	filled, pregnant
morpho-	shape, form
morul-	mulberry

WORD PART	MEANING
nat-	birth
oxy-	sharp, quick, rapid
para-	to bear
partur-	bring forth, give birth
sen-	old
-toc-	birth
umbil-	navel
zyg-	paired together, union

Use word parts given above or in previous chapters to form words that have the following definitions.

_______________ Development of shape or form

_______________ Study of the newborn

_______________ Surgical puncture of the amniotic sac

_______________ Quick childbirth; rapid labor

_______________ Ball of cells resembling a mulberry

Using the definitions of word parts given above or in previous chapters, define the following words.

Antenatal _______________

Pseudocyesis _______________

Dystocia _______________

Parturition _______________

Senescence _______________

Match each of the following definitions with the correct word.

_______ First pregnancy

_______ Excessive vomiting during pregnancy

_______ Cessation of milk secretion

_______ Excessive quantity of amniotic fluid

_______ Joining together of two cells

A. galactostasis

B. hyperemesis gravidarum

C. polyhydramnios

D. prima gravida

E. zygote

☞ Fun and Games

Each of the answers in this puzzle is a term from this chapter on development. Fill in the answers to the clues by using syllables from the list that is provided. The number of syllables in each word is indicated by the number in parentheses after the clue. The number of letters in each word is indicated by the number of spaces provided. All syllables in the list are to be used and no syllable is used more than once unless it is duplicated in the list.

A	E	LOS	O	TA
AD	E	MEN	O	TER
AM	FOR	MOR	O	TION
AT	GAN	NATE	ON	TOC
BLAST	GEN	NE	ON	TROPH
CAS	GEN	NES	OR	TRUM
CEN	GO	NI	OX	TU
CENCE	GOTE	NIX	PAR	U
CENCE	I	NU	PLA	VAGE
CHOR	IN	O	RI	VAL
CLEA	LA	O	SA	VER
CO	LA	O	SE	Y
E	LES	O	SIS	ZY

Product of fertilization (2) _ _ _ _ _ _

Early mitotic cell divisions after fertilization (2) _ _ _ _ _ _ _ _

Solid ball of blastomeres (3) _ _ _ _ _ _

Cells around the blastocele (3) _ _ _ _ _ _ _ _ _ _ _

Membrane around embryo/fetus (3) _ _ _ _ _ _

Extraembryonic membrane that contributes to placenta (3) _ _ _ _ _ _ _

Site of nutrient and gaseous exchange for fetus (3) _ _ _ _ _ _ _ _

Formation of body organs (6) _ _ _ _ _ _ _ _ _ _ _ _ _

Fine hair that covers fetus (3) _ _ _ _ _ _

Mixture of sebum and cells on fetal skin (6) _ _ _ _ _ _ _ _ _ _ _ _ _

Birth of an infant (4) _ _ _ _ _ _ _ _ _ _ _

Opening between R & L atria in fetus (6) _ _ _ _ _ _ _ _ _ _ _ _

Hormone that stimulates milk ejection (4) _ _ _ _ _ _ _ _

Secreted before milk production begins (3) _ _ _ _ _ _ _ _ _

Infant during first four weeks after birth (3) _ _ _ _ _ _ _

Period from puberty to adulthood (4) _ _ _ _ _ _ _ _ _ _ _

Period of old age (3) _ _ _ _ _ _ _ _ _ _

Substance that causes physical defects in embryo (4) _ _ _ _ _ _ _ _ _

Final Forty

Odd and Even: The final forty consists of two "odd and even" questions from each of the twenty chapters. Each question has a grouping of five (odd) items, four (even) of which are related in some way. You are to determine which is the "odd" item that does not belong with the rest and circle it. This will leave the four "even" items that are related.

1.	Ventral cavity	Cranial cavity	Thoracic cavity	Abdominal cavity	Pelvic cavity
2.	Frontal	Pectoral	Umbilical	Popliteal	Antecubital
3.	pH = 8	Lemon juice	Proton donor	Hydrogen ions	pH = 6
4.	Carbon	Maltose	Hydrogen	Carbohydrate	Nitrogen
5.	Lysosomes	Ribosomes	Mitochondria	Golgi apparatus	Nucleolus
6.	DNA	Ribosomes	Lysosomes	Proteins synthesis	RNA
7.	Adipose	Cartilage	Blood	Smooth muscle	Bone
8.	Epithelium	Meninges	Pleura	Mucous membrane	Serous membrane
9.	Papillary	Corneum	Basale	Granulosum	Lucidum
10.	Hair	Sebum	Holocrine	Sudoriferous	Sebaceous
11.	Zygomatic	Clavicle	Sphenoid	Sternum	Vertebrae
12.	Diarthrosis	Synovial membrane	Suture	Bursa	Joint capsule
13.	ATP	Creatine phosphate	Glucose	Fatty acids	Lactic acid
14.	Deltoid	Trapezius	Latissimus dorsi	Gracilis	Biceps brachii
15.	Neurilemma	Microglia	Axon	Myelin	Node of Ranvier
16.	Cerebrum	Cerebellum	Cranial nerves	Brainstem	Diencephalon
17.	Pupil	Cornea	Lens	Aqueous humor	Vitreous humor
18.	Tympanic membrane	Incus	Utricle	Stapes	Malleus
19.	TSH	FSH	ACTH	ADH	Prolactin
20.	Androgens	Epinephrine	Glucocorticoids	Aldosterone	Mineralocorticoids

21. Erythrocytes Hemoglobin Anucleate 5,000/mm^3 Biconcave discs

22. Gamma globulin Fibrinogen Prothrombin Platelets Calcium

23. Right atrium Superior vena cava Tricuspid valve Mitral valve Coronary sinus

24. QRS Ventricular filling 0.3 second Ventricular systole Blood ejection

25. Oxygenated blood Femoral vein Pulmonary vein Aorta Subclavian artery

26. Celiac artery Common hepatic artery Splenic artery Renal artery Left gastric aftery

27. Right side of the face Cisterna chyli Right leg Thoracic duct Left subclavian vein

28. Active immunity Vaccines Antiserum T-cells B-cells

29. Trachea Alveolus Bronchus Nasopharynx Bronchiole

30. Carbon dioxide Carbaminohemoglobin Oxyhemoglobin Bicarbonate ions Dissolved

31. Hydrochloric acid Intrinsic factor Enterokinase Pepsinogen Gastrin

32. Amylase Sucrase Maltase Lactase Lipase

33. Mitochondria Glycolysis Acetyl CoA Aerobic Citric acid cycle

34. Vitamin A Vitamin C Vitamin D Vitamin E Vitamin K

35. Glomerulus Renal corpuscle Proximal tubule Nephron loop Distal tubule

36. Extracellular fluid Sodium ions Interstitial fluid Plasma Potassium ions

37. Epididymis Ureter Ductus deferens Urethra Ejaculatory duct

38. Corona radiata Zona pellucida Primary follicle Vesicular follicle Secondary oocyte

39. Cleavage Blastomeres Morula Trophoblast Chorion

40. Fetal period Neonatal period Infancy Adolescence Adulthood

Glossary of Word Parts

a	without, lacking
ab	away from
acetabul	little cup
acoust	hearing
acro	extremity, point
act	motion
ad	toward
aden	gland
adip	fat
aero	air
af	toward
agglutin	clumping, sticking together
agon	assemble, gather together
al	pertaining to
albin	white
algia	pain
alkal	basic
alveol	tiny cavity
ambi	both
amnio	a fetal membrane, a lamb
amyl	starch
an	without, lack of
ana	up, apart
andr	male, maleness
angi	vessel
ante	before
anthrac	coal
anti	against
aort	lift up
apo	separation
appendicul	little attachment
arter	artery
arthr	joint
artic	joint
ary	pertaining to
ase	enzyme
asthenia	weakness
astro	star
atel	imperfect
ather	yellow fatty plaque
atri	entrance room
audi	to hear
auto	self
axill	armpit
balan	glans
betes	to go
bi	two
bili	bile, gall
blast	to form, sprout
blephar	eyelid
bol	throw, put
brachi	arm
brady	slow
bronchi	bronchi
bucc	cheek
burs	pouch
calcane	heel bone
carb/o	charcoal, coal, carbon
cardi	heart
carotid	put to sleep
carp	wrist
cata	down
caud	tail
celi	abdomen, belly
cente	to puncture surgically
cep	head
cephal	head
cer	wax
chole	gall, bile
chondr	cartilage
chromo	color
circum	around
clast	to break, destruction
cleav	to divide
cleido	clavicle
coagul	clotting
coch	snail
colpo	vagina
con	with, together
coni	dust
contra	against
corac	beak
corpor	body
cortic	outer region, cortex
cost	rib
cribr	sieve
cric	ring
crin	to secrete
crist	crest, ridge
crit	to separate
crypt	hidden
cusis	hearing
cusp	point
cutane	skin
cyesi	pregnancy
cyst	bladder, bag, hollow
cyt	cell
dactyl	finger or toe
de	down, away from
delt	triangle
dendr	tree

dent	tooth
derm	skin
di	two
dia	through
diastol	expand, separate
dipl	double
dips	thirst
dors	back
duct	movement
dur	hard
dys	difficult
ectasis	dilation
ecto	outer
ectomy	surgical removal
edem	to swell
ef	away from
ejacul	to shoot forth
elle	little, small
embol	stopper, wedge
emesis	vomit
emia	blood condition
encephal	within the head, brain
end	within, inner
enter	intestine
epi	upon, above
erg	work, energy
erythr	red
esthes	feeling
ethm	sieve
eu	good
ex	out of, away from
extra	outside, beyond
fasc	a band
fer	to carry
fibr	fiber
fic	make
fimb	fringe
flex	bend
follic	small bag
fove	pit
galacto	referring to milk
gangli	knot
gastr	stomach
gen	producing
genesis	to form, produce
ger	old age
gest	to carry, pregnancy
gingiv	gums
glia	glue
glob	globe
glomerul	little ball
gloss	tongue
gluc	sweet, sugar
glyc	sweet, sugar
gravid	filled, pregnant
gust	taste
gynec	female
hem	blood
hemi	half
hepat	liver
hetero	different
hex	six
hidr	sweat
hirsut	hairy
hist	tissue
holo	whole
homeo	alike, same
hydro	water
hyper	excessive, above
hypno	sleep
hypo	beneath, below
hyster	womb or uterus
ic	pertaining to
ichthy	scaly, dry
ide	pertaining to
immun	protection
in	neutral substance
infra	below
integ	a covering
inter	between
intra	within, inside
irid	iris
isch	deficiency
ism	process of
iso	equal, same
itis	inflammation
kary	nucleus
kerat	hard, horny tissue
kinesis	motion
kyph	hump
labi	lip
labyrinth	maze
lacrim	tear
lact	milk
laparo	flank, abdomen
laryng	larynx
lemm	peel, rind
leuk	white
lingu	tongue
lipo	fat
lith	stone

logy	study of, science of
lucid	clear, light
lun	moon shaped
lute	yellow
lymph	lymph
lys	to take apart
macro	large
macul	spot
mal	bad, poor
malacia	softening
mamm	breast
masset	chew
meat	passage
meg	large
melan	black
mening	membrane
meno	month
meso	middle
meta	between, change
metabol	change
metr	uterus
metr	measure
micro	small
micturit	to urinate
mnesia	memory
mono	one
morpho	shape, form
morul	mulberry
multi	many
myo, mys	muscle
naso	nose
nat	birth
neo	new
nephr	kidney
neur	nerve
noct	night
nulli	none
nutri	to nourish
ocul	eye
odont	tooth
oid	like, resembling
olfact	smell
olig	little, scanty
oma	tumor, swelling, mass
omion	shoulder
onychi	nail
oo	egg, ovum
oophor	ovary
op	eye
ophthalm	eye
orch	testicle
orexia	appetite
ortho	straight
ose	sugar
osis	condition of
oss	bone
oste	bone
ot	ear
oxy	sharp, quick, rapid
oxy	oxygen
pachy	thick
pan	all
para	beside
para	to bear
paresis	weakness
partur	bring forth, give birth
path	disease
pause	cessation
ped	foot
pelv	basin
penia	deficiency, lack of
pent	five
pept	to digest
peri	all around
pexy	fixation
phag	to eat, devour
phalange	closely knit row
pharyng	throat
phas	speech
phil	love, affinity for
phleb	vein
phob	hate, dislike
phon	voice
phragm	fence, partition
physi	nature, function
pin	pine cone
pino	to drink
plasm	matter
plasty	surgical repair
pleg	paralysis
plex	interweave, network
pnea	breathing
pneum	lung, air
poie	making
poly	many
post	after, behind
prandi	meal
pre	before, in front
presby	old
primi	first
pro	before
proct	rectum, anus
prostat	prostate
proto	first

proxim	nearest
pseudo	false
psych	mind
ptosis	prolapse, drooping
ptysis	spitting
pulmon	lung
pyel	renal pelvis
pyo	pus
pyr	fever, fire
quadri	four
ren	kidney
reti	network, lattice
retro	backward
rhin	nose
rhytido	wrinkles
rrhage	burst forth, flow
rrhaphy	suture
rrhea	flow or discharge
sacchar	sugar, sweet
salping	tube
sarco	flesh, muscle
scler	hard
scoli	curvature
seb	oil
semi	half
semin	seed
sen	old
sial	saliva
skelet	a dried, hard body
som	body
somn	sleep
spadias	an opening
sphen	wedge
sphygm	pulse
spir	breath
splen	spleen
squam	flattened, scale
sta	to control, staying
stalsis	contraction
sten	narrowing
ster	steroid
sterno	sternum
stomy	new opening
strat	layer
sub	below, under
sud	sweat
sulc	furrow, ditch
supra	above, beyond
sym	together
syn	together
systol	contraction
tachy	fast, rapid
terat	monster
test	eggshells, eggs
tetan	stiff
therm	temperature
thromb	clot
thym	thymus
thyr	shield
tion	act of, process of
toc	birth
tom	to cut
ton	solute strength
ton	tone, tension
tox	poison
tri	three
tripsy	crushing
trop	to change, influence
troph	nourish, develop
tuss	cough
tympan	drum
ul, ule	small, tiny
uln	elbow
ultra	beyond
um	presence of
umbil	navel
ungu	nail
uni	one
uria	urine condition
vago	twisted, wandering
valvu	valve
vas	vessel
ven	vein
ventilat	to fan or blow
verm	worm
vesic	bladder
viscer	internal organs
vita	life
vitre	glass
xer	dry
xiph	sword
y	process, condition
zyg	paired together, union